SO-ABA-372

National
Earthquake
Resilience

RESEARCH, IMPLEMENTATION, AND OUTREACH

Committee on National Earthquake Resilience—
Research, Implementation, and Outreach

Committee on Seismology and Geodynamics

Board on Earth Sciences and Resources

Division on Earth and Life Studies

NATIONAL RESEARCH COUNCIL
OF THE NATIONAL ACADEMIES

THE NATIONAL ACADEMIES PRESS
Washington, D.C.
www.nap.edu

THE NATIONAL ACADEMIES PRESS   500 Fifth Street, N.W.   Washington, DC 20001

NOTICE: The project that is the subject of this report was approved by the Governing Board of the National Research Council, whose members are drawn from the councils of the National Academy of Sciences, the National Academy of Engineering, and the Institute of Medicine. The members of the committee responsible for the report were chosen for their special competences and with regard for appropriate balance.

This study was supported by the National Institute of Standards and Technology under contract No. SB134106Z0011. The opinions, findings, and conclusions or recommendations contained in this document are those of the authors and do not necessarily reflect the views of the National Institute of Standards and Technology.

International Standard Book Number-13: 978-0-309-18677-3
International Standard Book Number-10: 0-309-18677-3
Library of Congress Control Number: 2011933648

Additional copies of this report are available from the National Academies Press, 500 Fifth Street, N.W., Lockbox 285, Washington, DC 20055; (800) 624-6242 or (202) 334-3313 (in the Washington metropolitan area); Internet www.nap.edu.

Cover: Cover design by Francesca Moghari. Seismogram images courtesy of iStockphoto LP.

THE NATIONAL ACADEMIES
Advisers to the Nation on Science, Engineering, and Medicine

The **National Academy of Sciences** is a private, nonprofit, self-perpetuating society of distinguished scholars engaged in scientific and engineering research, dedicated to the furtherance of science and technology and to their use for the general welfare. Upon the authority of the charter granted to it by the Congress in 1863, the Academy has a mandate that requires it to advise the federal government on scientific and technical matters. Dr. Ralph J. Cicerone is president of the National Academy of Sciences.

The **National Academy of Engineering** was established in 1964, under the charter of the National Academy of Sciences, as a parallel organization of outstanding engineers. It is autonomous in its administration and in the selection of its members, sharing with the National Academy of Sciences the responsibility for advising the federal government. The National Academy of Engineering also sponsors engineering programs aimed at meeting national needs, encourages education and research, and recognizes the superior achievements of engineers. Dr. Charles M. Vest is president of the National Academy of Engineering.

The **Institute of Medicine** was established in 1970 by the National Academy of Sciences to secure the services of eminent members of appropriate professions in the examination of policy matters pertaining to the health of the public. The Institute acts under the responsibility given to the National Academy of Sciences by its congressional charter to be an adviser to the federal government and, upon its own initiative, to identify issues of medical care, research, and education. Dr. Harvey V. Fineberg is president of the Institute of Medicine.

The **National Research Council** was organized by the National Academy of Sciences in 1916 to associate the broad community of science and technology with the Academy's purposes of furthering knowledge and advising the federal government. Functioning in accordance with general policies determined by the Academy, the Council has become the principal operating agency of both the National Academy of Sciences and the National Academy of Engineering in providing services to the government, the public, and the scientific and engineering communities. The Council is administered jointly by both Academies and the Institute of Medicine. Dr. Ralph J. Cicerone and Dr. Charles M. Vest are chair and vice chair, respectively, of the National Research Council.

www.national-academies.org

Preface

Earthquakes threaten much of the United States—damaging earthquakes struck Alaska in 1964 and 2002, California in 1857 and 1906, and the central Mississippi River Valley in 1811 and 1812. Moderate earthquakes causing substantial damage have repeatedly struck most of the western states as well as several mid-western and eastern states, e.g., South Carolina in 1886 and Massachusetts in 1755. The recent, disastrous, magnitude-9 earthquake that struck northern Japan demonstrates the threat that earthquakes pose, and the tragic impacts are especially striking because Japan is an acknowledged leader in implementing earthquake-resilient measures.[1] Moreover, the cascading nature of impacts—the earthquake causing a tsunami, cutting electrical power supplies, and stopping the pumps needed to cool nuclear reactors—demonstrates the potential complexity of an earthquake disaster. Such compound disasters can strike any earthquake-prone populated area.

Much can be done to mitigate the impact of earthquakes. Active fault zones and unstable ground can be avoided through wise land-use practices. Application of earthquake-resistant building codes and practices can reduce damage and casualties. Insurance and government assistance can facilitate recovery and ease economic impacts. And rapid response can save lives and restore essential services. Beyond these traditional approaches to reducing earthquake losses, there is a need for increased attention to the actions necessary for communities to rebound from an earthquake disaster.

---

[1] This tragedy occurred during report production, after the report had been completed and reviewed, so the committee was not able to include it in its analysis.

Recognizing the earthquake threat and the need to improve mitigation measures, Congress established the National Earthquake Hazards Reduction Program (NEHRP) in 1977 and has periodically reauthorized the program to the present time. NEHRP charges four federal agencies—the Federal Emergency Management Agency (FEMA), National Institute of Standards and Technology (NIST), National Science Foundation (NSF), and U.S. Geological Survey (USGS)—to advance knowledge of earthquake causes and effects and to develop and promulgate measures to reduce their impacts.

NIST, in its role as NEHRP lead agency, published a Strategic Plan for NEHRP in 2008 for the years 2009-2013, specifying the program's vision, mission, goals and objectives (NIST, 2008; summarized in Appendix A). In 2009, NIST requested that the National Research Council of the National Academies conduct a study, building on the Strategic Plan, to recommend a roadmap of national needs in research, knowledge transfer, implementation, and outreach to provide the tools to make the United States more earthquake resilient. Further, NIST requested that the roadmap use the results of a 2003 report by the Earthquake Engineering Research Institute titled *Securing Society Against Earthquake Losses—A Research and Outreach Plan in Earthquake Engineering* (EERI, 2003b; summarized in Appendix B). The EERI report includes cost projections for the program over a 20-year period, based on expert opinion, which NIST requested be updated and validated by our committee.

To carry out the study, the NRC established the Committee on Earthquake Resilience—Research, Implementation, and Outreach, an ad hoc committee under the Division on Earth and Life Studies. The committee membership includes experts from the full range of disciplines involved with earthquake risk mitigation. It met four times, including a workshop at the National Academies' Beckman Center in Irvine, California, which was attended by the committee members and about 40 invited participants, including representatives of the NEHRP agencies. The contributions of the participants informed the committee about key issues and concerns regarding NEHRP and contributed substantially to formulating the recommendations in this report.

Robert M. Hamilton
Chair

Acknowledgments

This report was greatly enhanced by those who made presentations to the committee at the public committee meetings and by the participants at the open workshop sponsored by the committee to gain community input—David Applegate, Walter Arabasz, Ralph Archuleta, Mark Benthien, Jonathan Bray, Arrietta Chakos, Mary Comerio, Reginald DesRoches, Andrea Donnellan, Leonardo Duenas-Osorio, Paul Earle, Richard Eisner, Ronald Eguchi, John Filson, Richard Fragaszy, Art Frankel, James Goltz, Ronald Hamburger, Jim Harris, Jack Hayes, Jon Heintz, Eric Holdeman, Doug Honegger, Richard Howe, Theresa Jefferson, Lucy Jones, Ed Laatsch, Michael Lindell, Nicolas Luco, Steven Mahin, Mike Mahoney, Peter May, Dick McCarthy, David Mendonça, Dennis Mileti, Robert Olson, Joy Pauschke, Chris Poland, Woody Savage, Hope Seligson, Kimberley Shoaf, Paul Somerville, Shyam Sunder, Kathleen Tierney, Susan Tubbesing, John Vidale, Yumei Wang, Gary Webb, Dennis Wenger, Sharon Wood, and Eva Zanzerkia. The presentations and discussions at these meetings provided invaluable input and context for the committee's deliberations.

This report has been reviewed in draft form by individuals chosen for their diverse perspectives and technical expertise, in accordance with procedures approved by the NRC's Report Review Committee. The purpose of this independent review is to provide candid and critical comments that will assist the institution in making its published report as sound as possible and to ensure that the report meets institutional standards for objectivity, evidence, and responsiveness to the study charge. The review comments and draft manuscript remain confidential to protect the integrity of the deliberative process. We wish to thank the following individuals for their participation in the review of this report:

John T. Christian, Independent Consultant, Waltham, Massachusetts
Lloyd S. Cluff, Pacific Gas and Electric Company, San Francisco, California
James H. Dieterich, University of California, Riverside
Carl A. Maida, University of California, Los Angeles
Chris D. Poland, Degenkolb Engineers, San Francisco, California
Barbara A. Romanowicz, University of California, Berkeley
Hope A. Seligson, MMI Engineering, Huntington Beach, California

Although the reviewers listed above have provided many constructive comments and suggestions, they were not asked to endorse the conclusions or recommendations nor did they see the final draft of the report before its release. The review of this report was overseen by Ross B. Corotis, Department of Civil, Environmental and Architectural Engineering, University of Colorado at Boulder, and Warren M. Washington, National Center for Atmospheric Research, Boulder, Colorado. Appointed by the National Research Council, they were responsible for making certain that an independent examination of this report was carried out in accordance with institutional procedures and that all review comments were carefully considered. Responsibility for the final content of this report rests entirely with the authoring committee and the institution.

Contents

Summary

The United States will certainly be subject to damaging earthquakes in the future, and some of those earthquakes will occur in highly populated and vulnerable areas. Just as Hurricane Katrina tragically demonstrated for hurricane events, coping with moderate earthquakes is not a reliable indicator of preparedness for a major earthquake in a populated area. This report presents a roadmap for increasing our national resilience to earthquakes, including the infrequent—but inevitable—Katrina-like earthquake events.

The United States has not experienced a great[1] earthquake since 1964, when Alaska was struck by a magnitude-9.2 event, and the damage in Alaska was relatively light because of the sparse population. The 1906 San Francisco earthquake was the most recent truly devastating U.S. shock, because recent destructive earthquakes have been only moderate to strong in size. Consequently, a sense has developed that the country can cope effectively with the earthquake threat and is, in fact, "resilient." However, coping with moderate events may not be a true indicator of preparedness for a great one. One means to understand the potential effects from major earthquakes is to use scenarios, where communities simulate the effects and responses to a specified earthquake. Analysis of the 2008 ShakeOut scenario in California (Jones et al., 2008), which involved more than 5,000

---

[1] Damaging effects from earthquakes reflect not only the earthquake magnitude, but also ground motion as measured by velocity, acceleration, frequency, and shaking duration. U.S. Geological Survey definitions of earthquake magnitude classes are "great" =M≥8; "major" M=7-7.9; "strong" M=6-6.9; "moderate" M=5-5.9; etc. See earthquake.usgs.gov/learn/faq/?faqID=24.

emergency responders and the participation of more than 5.5 million citizens, indicated that the magnitude-7.8 scenario earthquake would have resulted in an estimated 1,800 fatalities, $113 billion in damages to buildings and lifelines, and nearly $70 billion in business interruption. Such an earthquake would clearly have a major effect on the nation as a whole, emphasizing the need to develop the capacity to reduce such effects—to increase our national earthquake resilience.

The National Earthquake Hazards Reduction Program (NEHRP) is the multi-agency program mandated by Congress to undertake activities to reduce the effects of future earthquakes in the United States. NEHRP was initially authorized by Congress in 1977 and subsequently reauthorized on 2- to 5-year intervals. The four federal agencies with funding authorizations and legislatively mandated responsibilities for NEHRP activities are the Federal Emergency Management Agency (FEMA), the National Institute of Standards and Technology (NIST), the National Science Foundation (NSF), and the U.S. Geological Survey (USGS). In 2009, NEHRP funding was $129.7 million, allocated to the USGS ($61.2 million), NSF ($55.3 million), FEMA ($9.1 million), and NIST ($4.1 million) (NIST, 2008). In 2008, the NEHRP agencies developed a Strategic Plan with the aim of providing a sound basis for future activities. The plan is focused on 14 objectives that are grouped into three major goals: to improve understanding of earthquake processes and impacts; to develop cost-effective measures to reduce earthquake impacts on individuals, the built environment, and society-at-large; and to improve the earthquake resilience of communities nationwide.

NIST—the lead NEHRP agency—commissioned the National Research Council (NRC) to develop a roadmap for earthquake hazard and risk reduction in the United States that would be based on the goals and objectives for achieving national earthquake resilience described in the 2008 NEHRP Strategic Plan. The NRC committee was directed to assess the activities, and their costs, that would be required for the nation to achieve earthquake resilience in 20 years. The charge to the committee recognized that there would be a requirement for some sustained activities under the NEHRP program after this 20-year period (see full statement of task in Chapter 1, Box 1.2).

DEFINING EARTHQUAKE RESILIENCE

A critical requirement for achieving national earthquake resilience is, of course, an understanding of what constitutes earthquake resilience. In this report, we have interpreted resilience broadly so that it incorporates engineering/science (physical), social/economic (behavioral), and institutional (governing) dimensions. Resilience is also interpreted to encompass both pre- and post-disaster actions that, in combination, will enhance the robustness and the capabilities of all earthquake-vulnerable regions of our nation to function adequately following damaging earthquakes. The committee is also cognizant that it is cost-prohibitive to achieve a completely seismically resistant nation. Instead, we see our mission as helping set performance targets for improving the nation's seismic resilience over the next 20 years and, in turn, developing a more detailed road map and program priorities for NEHRP. With these considerations in mind, the committee recommends that NEHRP adopt the following working definition for "national earthquake resilience":

> **A disaster-resilient nation is one in which its communities, through mitigation and pre-disaster preparation, develop the adaptive capacity to maintain important community functions and recover quickly when major disasters occur.**

ELEMENTS AND COSTS OF A RESILIENCE ROADMAP

The committee set out to build on the 2008 NEHRP Strategic Plan by specifying focused activities that would further implementation of the plan and provide the basis for a more earthquake-resilient nation. In the end, 18 tasks were identified, ranging from basic research to community-oriented applications, which, in our view, comprise a "roadmap" for furthering NEHRP goals and implementing the Strategic Plan. The tasks generally cross cut the goals and objectives described in the Strategic Plan because they are formulated as coherent activities that span from knowledge building to implementation.

> **The committee endorses the 2008 NEHRP Strategic Plan, and identifies 18 specific task elements required to implement that plan and materially improve national earthquake resilience.**

In estimating costs to implement the roadmap, the committee recognizes that there is a high degree of variability among the 18 tasks—some are under way or are in the process of being implemented, whereas others are only at the conceptual stage. Costing each task required a thorough analysis to determine scope, implementation steps, and linkages or overlaps with other tasks. For some of the tasks, the necessary analysis had already been completed in workshops or other venues, and realistic cost estimates were available as input to the committee (see Appendix E for cost estimate details). For other tasks, the committee had to rely on its own expert opinion, in which case implementing the task may require some degree of additional detailed analysis. In summary, the annualized cost for the first 5 years of the roadmap for national earthquake resilience presented here is $306.5 million/year (2009$), summarized in Table S.1 and made up of the following tasks:

1.   **Physics of Earthquake Processes.** Conduct additional research to advance the understanding of earthquake phenomena and earthquake generation processes and to improve the predictive capabilities of earthquake science; 5-year annualized cost of $27 million/year, for a total 20-year cost of $585 million.

2.   **Advanced National Seismic System.** Complete deployment of the remaining 75 percent of the Advanced National Seismic System; 5-year annualized cost of $66.8 million/year, for a total 20-year cost of $1.3 billion. On-going operations and maintenance costs after the initial 20-year period of $50 million/year.

3.   **Earthquake Early Warning.** Evaluation, testing, and deployment of earthquake early warning systems; 5-year annualized cost of $20.6 million/year, for a total 20-year cost of $283 million.

4.   **National Seismic Hazard Model.** Complete the national coverage of seismic hazard maps and create urban seismic hazard maps and seismic risk maps for at-risk communities; 5-year annualized cost of $50.1 million/year, for a total 20-year cost of $946.5 million.

5.   **Operational Earthquake Forecasting.** Develop and implement operational earthquake forecasting, in coordination with state and local agencies; 5-year annualized cost of $5 million/year, for a total 20-year cost of $85 million. On-going operations and maintenance costs after the initial 20-year period are unknown.

6.   **Earthquake Scenarios.** Develop scenarios that integrate earth science, engineering, and social science information so that communities can visualize earthquake and tsunami impacts and mitigate potential effects; 5-year annualized cost of $10 million/year, for a total 20-year cost of $200 million.

7.   **Earthquake Risk Assessments and Applications.** Integrate science, engineering, and social science information in an advanced GIS-based loss estimation platform to improve earthquake risk assessments and loss estimations; 5-year annualized cost of $5 million/year, for a total 20-year cost of $100 million.

8.   **Post-earthquake Social Science Response and Recovery Research.** Document and model the mix of expected and improvised emergency response and recovery activities and outcomes to improve pre-disaster mitigation and preparedness practices at household, organizational, community, and regional levels; 5-year annualized cost of $2.3 million/year, reviewed after the initial 5-years.

9.   **Post-earthquake Information Management.** Capture, distill, and disseminate information about the geological, structural, institutional,

TABLE S.1 Compilation of Cost Estimates by Task, in Millions of Dollars (all figures are 2009 dollars).

Task	Annualized Costs (av.) Years 1-5 ($)	Total Cost Years 1-5 ($)	Total Cost Years 6-20 ($)	Total Cost ($)
1. Physics of Earthquake Processes	27	135	450	585
2. Advanced National Seismic System (ANSS)[a]	66.8	334	1,002	1,336
3. Earthquake Early Warning	20.6	103	180	283
4. National Seismic Hazard Model	50.1	250.5	696	946.5
5. Operational Earthquake Forecasting	5	25	60	85
6. Earthquake Scenarios	10	50	150	200
7. Earthquake Risk Assessments and Applications	5	25	75	100
8. Post-earthquake Social Science Response and Recovery Research	2.3	11.5	TBD[b]	TBD[b]
9. Post-earthquake Information Management	1	4.8	9.8	14.6
10. Socioeconomic Research on Hazard Mitigation and Recovery	3	15	45	60
11. Observatory Network on Community Resilience and Vulnerability	2.9	14.5	42.8	57.3
12. Physics-based Simulations of Earthquake Damage and Loss	6	30	90	120
13. Techniques for Evaluation and Retrofit of Existing Buildings	22.9	114.5	429.1	543.6
14. Performance-based Earthquake Engineering for Buildings	46.7	233.7	657.8	891.5
15. Guidelines for Earthquake-Resilient Lifelines Systems	5	25	75	100
16. Next Generation Sustainable Materials, Components, and Systems	8.2	40.8	293.6	334.4
17. Knowledge, Tools, and Technology Transfer to Public and Private Practice	8.4	42	126	168
18. Earthquake-Resilient Communities and Regional Demonstration Projects	15.6	78	923	1,001
TOTAL	**306.5**	**1,532.3**	**5,305.1**	**6,837.4**

[a] Does not include support for geodetic monitoring or geodetic networks.
[b] Funding during the remaining 15 years of the plan would be based on a performance review after 5 years.

and socioeconomic impacts of specific earthquakes, as well as post-disaster response, and create and maintain a repository for post-earthquake reconnaissance data; 5-year annualized cost of $1 million/year, for a total 20-year cost of $14.6 million. On-going operations and maintenance costs after the initial 20-year period are unknown, but are likely to be small.

10. **Socioeconomic Research on Hazard Mitigation and Recovery.** Support basic and applied research in the social sciences to examine individual and organizational motivations to promote resilience, the feasibility and cost of resilience actions, and the removal of barriers to successful implementation; 5-year annualized cost of $3 million/year, for a total 20-year cost of $60 million.

11. **Observatory Network on Community Resilience and Vulnerability.** Establish an observatory network to measure, monitor, and model the disaster vulnerability and resilience of communities, with a focus on resilience and vulnerability; risk assessment, perception, and management strategies; mitigation activities; and reconstruction and recovery; of 5-year annualized cost $2.9 million/year, for a total 20-year cost of $57.3 million. On-going operations and maintenance costs after the initial 20-year period are unknown.

12. **Physics-based Simulations of Earthquake Damage and Loss.** Integrate knowledge gained in Tasks 1, 13, 14, and 16 to enable robust, fully coupled simulations of fault rupture, seismic wave propagation through bedrock, and soil-structure response to compute reliable estimates of financial loss, business interruption, and casualties; 5-year annualized cost of $6 million/year, for a total 20-year cost of $120 million.

13. **Techniques for Evaluation and Retrofit of Existing Buildings.** Develop analytical methods that predict the response of existing buildings with known levels of reliability based on integrated laboratory research and numerical simulations, and improve consensus standards for seismic evaluation and rehabilitation; 5-year annualized cost of $22.9 million/year, for a total 20-year cost of $543.6 million.

14. **Performance-based Earthquake Engineering for Buildings.** Advance performance-based earthquake engineering knowledge and develop implementation tools to improve design practice, inform decision-makers, and revise codes and standards for buildings, lifelines, and geo-structures; 5-year annualized cost of $46.7 million/year, for a total 20-year cost of $891.5 million.

15. **Guidelines for Earthquake-Resilient Lifeline Systems.** Conduct lifeline-focused collaborative research to better characterize infrastructure network vulnerability and resilience as the basis for the systematic review and updating of existing lifelines standards and guidelines, with targeted pilot programs and demonstration projects; 5-year annualized cost of $5 million/year, for a total 20-year cost of $100 million.

16. **Next Generation Sustainable Materials, Components, and Systems.** Develop and deploy new high-performance materials, components, and framing systems that are green and/or adaptive; 5-year annualized cost is $8.2 million/year, for a total 20-year cost of $334.4 million.

17. **Knowledge, Tools, and Technology Transfer to Public and Private Practice.** Initiate a program to encourage and coordinate technology transfer across the NEHRP domain to ensure the deployment of state-of-the-art mitigation techniques across the nation, particularly in regions of moderate seismic hazard; 5-year annualized cost of $8.4 million/year, for a total 20-year cost of $168 million.

18. **Earthquake-Resilient Community and Regional Demonstration Projects.** Support and guide community-based earthquake resiliency pilot projects to apply NEHRP-generated and other knowledge to improve awareness, reduce risk, and improve emergency preparedness and recovery capacity; 5-year annualized cost of $15.6 million/year, for a total 20-year cost of $1 billion.

TIMING OF ROADMAP COMPONENTS

The committee recommends that all the tasks identified here be initiated immediately, contingent on the availability of funds, and suggests that such an approach would represent an appropriate balance between practical activities to enhance national earthquake resilience and the research that is needed to provide a sound basis for such activities. However, at a lower component level within individual tasks, there are some elements that should be implemented and/or initiated immediately whereas others will have to await the results of earlier activities. Sequencing information and detailed cost breakdowns are listed for several tasks in Appendix E. The committee also notes that the two "observatory" elements of the roadmap, Task 2 and Task 11, will—or do at present—provide fundamental information to be used by numerous other tasks.

EARTHQUAKE RESILIENCE AND AGENCY COORDINATION

The four NEHRP agencies, although comprising a critical core group for building earthquake knowledge, constitute only part of the national research and application enterprise on which earthquake resilience is based. In the applications area, virtually every agency that builds or operates facilities contributes to the goals of NEHRP by adopting practices or codes to reduce earthquake impacts. These agencies include the U.S. Army Corps of Engineers and the Departments of Transportation, Energy, and Housing and Urban Development. Beyond the role of the federal agencies, government agencies at all levels similarly play a critical role in the application of earthquake knowledge, as does the private sector, especially in the area of building design. Altogether, the contributors to reducing earthquake losses constitute a complex enterprise that goes far beyond the scope of NEHRP. But NEHRP provides an important focus for this far-flung endeavor. The committee considers that an analysis to determine whether coordination among *all* organizations that contribute to NEHRP could be improved would be useful and timely.

IMPLEMENTING NEHRP KNOWLEDGE

Most critical decisions that reduce earthquake vulnerability and manage earthquake risk are made in the private sector by individuals and companies. The information provided by NEHRP, if made available in an understandable format and accompanied by diffusion processes, can greatly assist citizens in their decision-making. For example, maps of active faults, unstable ground, and historic seismicity can influence where people choose to live, and maps of relative ground shaking can guide building design.

NEHRP will have accomplished its fundamental purpose—an earthquake-resilient nation—when those responsible for earthquake risk and for managing the consequences of earthquake events use the knowledge and services created by NEHRP and other related endeavors to make our communities more earthquake resilient. Increasing resiliency requires awareness of earthquake risk, knowing what to do to address that risk, and doing it. But providing information is not enough to achieve resilience—the diffusion of NEHRP knowledge and implementation of that knowledge are necessary corollaries. Successfully diffusing NEHRP knowledge into communities and among the earthquake professionals, state and local government officials, building owners, lifeline operators, and others who have the responsibility for how buildings, systems, and institutions respond to and recover from earthquakes, will require a dedicated and strategic effort. This diffusion role reflects the limited authority that resides with federal agencies in addressing the earthquake threat. Local and state governments have responsibility for public safety and welfare, including powers to regulate land use to avoid hazards, establish and enforce building codes to withstand earthquake forces, provide warnings to threatened communities, and respond to an event. The goals and objectives of NEHRP are aimed at supporting and facilitating measures to improve resilience through private owners and businesses, and supporting local and state agencies in carrying out their duties. Although implementing NEHRP knowledge should move ahead expeditiously, it is also essential that the frontiers of knowledge be advanced in concert, requiring that improving understanding of the earthquake threat, reducing risk, and developing the processes to motivate implementation actions, should all be continuing endeavors.

1

Introduction

When a strong earthquake hits an urban area, structures collapse, people are injured or killed, infrastructure is disrupted, and business interruption begins. The immediate impacts caused by an earthquake can be devastating to a community, challenging it to launch rescue efforts, restore essential services, and initiate the process of recovery. The ability of a community to recover from such a disaster reflects its resilience, and it is the many factors that contribute to earthquake resilience that are the focus of this report. Specifically, we provide a roadmap for building community resilience within the context of the Strategic Plan of the National Earthquake Hazards Reduction Program (NEHRP), a program first authorized by Congress in 1977 to coordinate the efforts of four federal agencies—National Institute of Standards and Technology (NIST), Federal Emergency Management Agency (FEMA), National Science Foundation (NSF), and U.S. Geological Survey (USGS).

The three most recent earthquake disasters in the United States all occurred in California—in 1994 near Los Angeles at Northridge, in 1989 near San Francisco centered on Loma Prieta, and in 1971 near Los Angeles at San Fernando. In each earthquake, large buildings and major highways were heavily damaged or collapsed and the economic activity in the afflicted area was severely disrupted. Remarkably, despite the severity of damage, deaths numbered fewer than a hundred for each event. More-over, in a matter of days or weeks, these communities had restored many essential services or worked around major problems, completed rescue efforts, and economic activity—although impaired—had begun to recover. It could be argued that these communities were, in fact, quite resilient. But it should be emphasized that each of these earthquakes was only moderate to strong in size, less than magnitude-7, and that the impacted areas were limited in size. How well would these communities cope with a magnitude-8 earthquake? What lessons can be drawn from the resilience demonstrated for a moderate earthquake in preparing for a great one?

Perhaps experience in dealing with hurricane disasters would be instructive in this regard. In a typical year, a few destructive hurricanes make landfall in the United States. Most of them cause moderate structural damage, some flooding, limited disruption of services—usually loss of power—and within a few days, activity returns to near normal. However, when Hurricane Katrina struck the New Orleans region in 2005 and caused massive flooding and long-term evacuation of much of the population, the response capabilities were stretched beyond their limits. Few observers would argue that New Orleans, at least in the short term, was a resilient community in the face of that event.

Would an earthquake on the scale of the 1906 event in northern California or the 1857 event in southern California lead to a similar catastrophe? It is likely that an earthquake on the scale of these events in California would indeed lead to a catastrophe similar to hurricane Katrina, but of a significantly different nature. Flooding, of course, would not be the main hazard, but substantial casualties, collapse of structures, fires, and economic disruption could be of great consequence. Similarly, what would happen if there were to be a repeat of the New Madrid earthquakes of 1811-1812, in view of the vulnerability of the many bridges and chemical facilities in the region and the substantial barge traffic on the Mississippi River? Or, consider the impact if an earthquake like the 1886 Charleston tremor struck in other areas in the central or eastern United States, where earthquake-prone, unreinforced masonry structures abound and earthquake preparedness is not a prime concern? The resilience of communities and regions, and the steps—or roadmap—that could be taken to ensure that areas at risk become earthquake resilient, are the subject of this report.

EARTHQUAKE RISK AND HAZARD

Earthquakes proceed as cascades, in which the primary effects of faulting and ground shaking induce secondary effects such as landslides, liquefaction, and tsunami, which in turn set off destructive processes within the built environment such as fires and dam failures (NRC, 2003). The socioeconomic effects of large earthquakes can reverberate for decades.

The *seismic hazard* for a specified site is a probabilistic forecast of how intense the earthquake effects will be at that site. In contrast, *seismic risk* is a probabilistic forecast of the damage to society that will be caused by earthquakes, usually measured in terms of casualties and economic losses in a specified area integrated over the post-earthquake period. Risk depends on the hazard, but it is compounded by a community's *exposure*—its population and the extent and density of its built environment—as well as the *fragility* of its built environment, population, and socioeconomic systems to seismic hazards. Exposure and fragility contribute to *vulnerability*. Risk is lowered by *resiliency*, the measure of how efficiently and how quickly a community can recover from earthquake damage.

Risk analysis seeks to quantify the risk equation in a framework that allows the impact of political policies and economic investments to be evaluated, to inform the decision-making processes that contribute to risk reduction. Risk quantification is a difficult problem, because it requires detailed knowledge of the natural and the built environments, as well as an understanding of both earthquake and human behaviors. Moreover, national risk is a dynamic concept because of the exponential rise in the urban exposure to seismic hazards (EERI, 2003b)—calculating risk involves predictions of highly uncertain demographic trends.

Estimating Losses from Earthquakes

The synoptic earthquake risk studies needed for policy formulation are the responsibility of NEHRP. These studies can take the form of deterministic or scenario studies where the effects of a single earthquake are modeled, or probabilistic studies that weight the effects from a number of different earthquake scenarios by the annual likelihood of their occurrence. The consequences are measured in terms of dollars of damage, fatalities, injuries, tons of debris generated, ecological damage, etc. The exposure period may be defined as the design lifetime of a building or some other period of interest (e.g., 50 years). Typically, seismic risk estimates are presented in terms of an exceedance probability (EP) curve (Kunreuther et al., 2004), which shows the probability that specific parameters will equal or exceed specified values (Figure 1.1). On this figure, a loss estimate calculated for a specific scenario earthquake is represented by a horizontal slice through the EP curve, while estimates of annualized losses from earthquakes are portrayed by the area under the EP curve.

The 2008 Great California ShakeOut exercise in southern California is an example of a scenario study that describes what would happen during and after a magnitude-7.8 earthquake on the southernmost 300 km of the San Andreas Fault (Figure 1.2), a plausible event on the fault that is most likely to produce a major earthquake. Analysis of the 2008 ShakeOut scenario, which involved more than 5,000 emergency responders and the participation of more than 5.5 million citizens, indicated that the scenario earthquake would have resulted in an estimated 1,800 fatalities, $113 billion in damages to buildings and lifelines, and nearly $70 billion in busi-

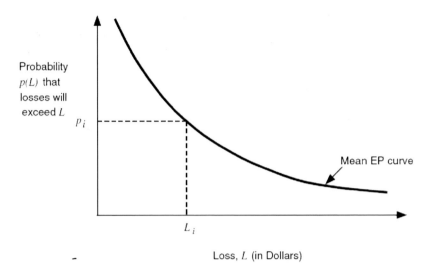

FIGURE 1.1 Sample mean EP curve, showing that for a specified event the probability of insured losses exceeding L_i is given by p_i. SOURCE: Kunreuther et al. (2004).

ness interruption (Jones et al., 2008; Rose et al., in press). The broad areal extent and long duration of water service outages was the main contributor to business interruption losses. Moreover, the scenario is essentially a compound event like Hurricane Katrina, with the projected urban fires caused by gas main breaks and other types of induced accidents projected to cause $40 billion of the property damage and more than $22 billion of the business interruption. Devastating fires occurred in the wake of the 1906 San Francisco, 1923 Tokyo, and 1995 Kobe earthquakes.

Loss estimates have been published for a range of earthquake scenarios based on historic events—e.g., the 1906 San Francisco earthquake (Kircher et al., 2006); the 1811/1812 New Madrid earthquakes (Elnashai et al., 2009); and the magnitude-9 Cascadia subduction earthquake of 1700 (CREW, 2005)—or inferred from geologic data that show the magnitudes and locations of prehistoric fault ruptures (e.g., the Puente Hills blind thrust that runs beneath central Los Angeles; Field et al., 2005). In all cases, the results from such estimates are staggering, with economic losses that run into the hundreds of billions of dollars.

FEMA's latest estimate of Annualized Earthquake Loss (AEL) for the nation (FEMA, 2008) is an example of a probabilistic study—an estimate of national earthquake risk that used HAZUS-MH software (Box 1.1) together with input from Census 2000 data and the 2002 USGS National Seismic Hazard Map. The current AEL estimate of $5.3 billion (2005$)

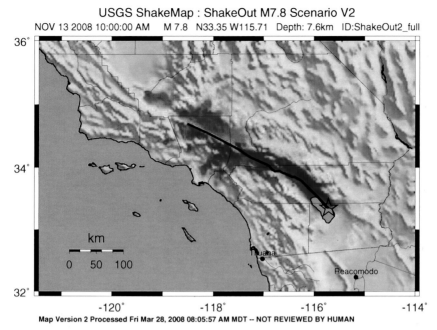

FIGURE 1.2 A "ShakeMap" representing the shaking produced by the scenario earthquake on which the Great California ShakeOut was based. The colors represent the Modified Mercalli Intensity, with warmer colors representing areas of greater damage. SOURCE: USGS. Available at earthquake.usgs.gov/earthquakes/shakemap/sc/shake/ShakeOut2_full_se/.

reflects building-related direct economic losses including damage to buildings and their contents, commercial inventories, as well as damaged building-related income losses (e.g., wage losses, relocation costs, rental income losses, etc.), but does not include indirect economic losses or losses to lifeline systems. For comparison, the Earthquake Engineering Research Institute (EERI) (2003b) extrapolated the FEMA (2001) estimate of AEL ($4.4 billion) for residential and commercial building-related direct economic losses by a factor of 2.5 to include indirect economic losses, the social costs of death and injury, as well as direct and indirect losses to the

---

BOX 1.1
HAZUS®—Risk Metrics for NEHRP

The ability to monitor and compare seismic risk across states and regions is critical to the management of NEHRP. At the state and local level, an understanding of seismic risk is important for planning and for evaluating costs and benefits associated with building codes, as well as a variety of other prevention measures. HAZUS is Geographic Information System (GIS) software for earthquake loss estimation that was developed by FEMA in cooperation with the National Institute of Building Sciences (NIBS). HAZUS-MH (Hazards U.S.-Multi-Hazard) was released in 2003 to include wind and flood hazards in addition to the earthquake hazards that were the subject of the 1997 and 1999 HAZUS releases. Successive HAZUS maintenance releases (MR) have been made available by FEMA since the initial HAZUS-MH MR-1 release; the latest version, HAZUS-MH MR-5, was released in December 2010.

Annualized Earthquake Loss (AEL) is the estimated long-term average of earthquake losses in any given year for a specific location. Studies by FEMA based on the 1990 and 2000 censuses provide two "snapshots" of seismic risk in the United States (FEMA, 2001, 2008). These studies, together with an earlier analysis of the 1970 census by Petak and Atkisson (1982), show that the estimated national AEL increased from $781 million (1970$) to $4.7 billion (2000$)—or by about 40 percent—over four decades (Figure 1.3). All three studies used building-related direct economic losses and included structural and nonstructural replacement costs, contents damage, business inventory losses, and direct business interruption losses.

---

industrial, manufacturing, transportation, and utility sectors to arrive at an annual average financial loss in excess of $10 billion.

Although the need to address earthquake risk is now accepted in many communities, the ability to identify and act on specific hazard and risk issues can be improved by reducing the uncertainties in the risk equation. Large ranges in loss estimates generally stem from two types of uncertainty—the natural variability assigned to earthquake processes (*aleatory uncertainty*), as well as a lack of knowledge of the true hazards and risks involved (*epistemic uncertainty*). Uncertainties are associated with the methodologies, the assumptions, and databases used to estimate the ground motions and building inventories, the modeling of building responses, and the correlation of expected economic and social losses to the estimated physical damages.

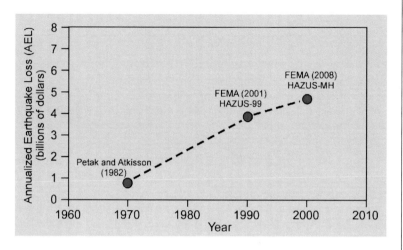

FIGURE 1.3 Growth of seismic risk in the United States. Annualized Earthquake Loss (AEL) estimates are shown for the census year on which the estimate is based, in census year dollars. Estimate for 1970 census from Petak and Atkinson (1982); HAZUS-99 estimate for 1990 census from FEMA (2001); and HAZUS-MH estimate for 2000 census from FEMA (2008). Consumer Price Index (CPI) dollar adjustments based on CPI inflation calculator (see data.bls.gov/cgi-bin/cpicalc.pl).

Comparison of published risk estimates reveals the sensitivity of such estimates to varying inputs, such as soil types and ground motion attenuation models, or building stock inventories and damage calculations. The basic earth science and geotechnical research and data that the NEHRP agencies provide to communities help to reduce these types of epistemic uncertainty, whereas an understanding of the intrinsic aleatory uncertainty is achieved through scientific research into the processes that cause earthquakes. Accurate loss estimation models increase public confidence in making seismic risk management decisions. Until the uncertainties surrounding the EP curve in Figure 1.1 are reduced, there will be either unnecessary or insufficient emergency response planning and mitigation because the experts in these areas will be unable to inform decision-makers of the probabilities and potential outcomes with an appropriate degree of confidence (NRC, 2006a). Information about new and rehabilitated buildings and infrastructure, coupled with improved seismic hazard maps, can allow policy-makers to track incremental reductions in risk and improvements in safety through earthquake mitigation programs (NRC, 2006b).

NEHRP ACCOMPLISHMENTS—THE PAST 30 YEARS

In its 30 years of existence, NEHRP has provided a focused, coordinated effort toward developing a knowledge base for addressing the earthquake threat. The following summary of specific accomplishments from the earth sciences and engineering fields are based on the 2008 NEHRP Strategic Plan (NIST, 2008):

- **Improved understanding of earthquake processes.** Basic research and earthquake monitoring have significantly advanced the understanding of the geologic processes that cause earthquakes, the characteristics of earthquake faults, the nature of seismicity, and the propagation of seismic waves. This understanding has been incorporated into seismic hazard assessments, earthquake potential assessments, building codes and design criteria, rapid assessments of earthquake impacts, and scenarios for risk mitigation and response planning.
- **Improved earthquake hazard assessment.** Improvements in the National Seismic Hazard Maps have been developed through a scientifically defensible and repeatable process that involves peer input and review at regional and national levels by expert and user communities. Once based on six broad zones, they now are based on a grid of seismic hazard assessments at some 150,000 sites throughout the country. The new maps, first developed in 1996, are periodically updated and form the basis for the Design Ground Motion Maps used in the NEHRP Recommended Provisions for Seismic Regulations for New Buildings and Other Structures, the foundation for the seismic elements of model building codes.
- **Improved earthquake risk assessment.** Development of earthquake hazard- and risk-assessment techniques for use throughout the United States has improved awareness of earthquake impacts on communities. NEHRP funds have supported the development and continued refinement of HAZUS-MH. The successful NEHRP-supported integration of earthquake risk-assessment and loss-estimation methodologies with earthquake hazard assessments and notifications has provided significant benefits for both emergency response and community planning. Moreover, major advances in risk assessment and hazard loss estimation beyond what could be included in a software package for general users were developed by the three NSF-supported earthquake engineering centers.

- **Improved earthquake safety in design and construction.** Earthquake safety in new buildings has been greatly improved through the adoption, in whole or in part, of earthquake-resistant national model building codes by state and local governments in all 50 states. Development of advanced earthquake engineering technologies for use in design and construction has greatly improved the cost-effectiveness of earthquake-resistant design and construction while giving options with predicted decision consequences. These techniques include new methods for reducing the seismic risk associated with nonstructural components, base isolation methods for dissipating seismic energy in buildings, and performance-based design approaches.
- **Improved earthquake safety for existing buildings.** NEHRP-led research, development of engineering guidelines, and implementation activities associated with existing buildings have led to the first generation of consensus-based national standards for evaluating and rehabilitating existing buildings. This work provided the basis for two American Society of Civil Engineers (ASCE) standards documents: ASCE 31 (Seismic Evaluation of Existing Buildings) and ASCE 41 (Seismic Rehabilitation of Existing Buildings).
- **Development of partnerships for public awareness and earthquake mitigation.** NEHRP has developed and sustained partnerships with state and local governments, professional groups, and multi-state earthquake consortia to improve public awareness of the earthquake threat and support the development of sound earthquake mitigation policies.
- **Improved development and dissemination of earthquake information.** There is now a greatly increased body of earthquake-related information available to public- and private-sector officials and the general public. This comes through effective documentation, earthquake response exercises, learning-from-earthquake activities, publications on earthquake safety, training, education, and information on general earthquake phenomena and means to reduce their impact. Millions of earthquake preparedness handbooks have been delivered to at-risk populations, and many of these handbooks have been translated from English into languages most easily understood by large sectors of the population. NEHRP now maintains a website[1] that provides information on the program and communicates regularly with the earthquake professional community through the monthly electronic newsletter, Seismic Waves.
- **Improved notification of earthquakes.** The USGS National Earthquake Information Center and regional networks, all elements of the Advanced National Seismic System (ANSS), now provide earthquake

---

[1] See www.nehrp.gov.

alerts describing a magnitude and location within a few minutes after an earthquake. The USGS PAGER system[2] provides estimates of the number of people and the names of cities exposed to shaking, with corresponding levels of impact shown by the Modified Mercalli Intensity scale and estimates of the number of fatalities and economic loss, following significant earthquakes worldwide (Figure 1.4). When coupled with graphic ShakeMaps[3] showing the distribution and severity of ground shaking (e.g., Chapter 3, Figure 3.2), this information is essential for effective emergency response, infrastructure management, and recovery planning.

• **Expanded training and education of earthquake professionals.** Thousands of graduates of U.S. colleges and universities have benefited from their involvement and experiences with NEHRP-supported research projects and training activities. Those graduates now form the nucleus of America's earthquake professional community.

• **Development of advanced data collection and research facilities.** NEHRP took the lead in developing ANSS and the George E. Brown, Jr. Network for Earthquake Engineering Simulation (NEES). Through these initiatives, NEES now forms a national infrastructure for testing geotechnical, structural, and nonstructural systems, and once completed, ANSS will provide a comprehensive, nationwide system for monitoring seismicity and collecting data on earthquake shaking on the ground and in structures. NEHRP also has participated in the development of the Global Seismographic Network to provide data on seismic events worldwide.

As well as this list of important accomplishments cited in the 2008 NEHRP Strategic Plan, the following range of NEHRP accomplishments in the social science arena were described in NRC (2006a):

• **Development of a comparative research framework.** Largely supported by NEHRP, over the past three decades social scientists increasingly have placed the study of earthquakes within a comparative framework that includes other natural, technological, and willful events. This evolving framework calls for the integration of hazards and disaster research within the social sciences and among social science, natural science, and engineering disciplines.

• **Documentation of community and regional vulnerability to earthquakes and other natural hazards.** Under NEHRP sponsorship, social science knowledge has expanded greatly in terms of data on community and regional exposure and vulnerability to earthquakes and other natural hazards, such that the foundation has been established for devel-

---

[2] See earthquake.usgs.gov/earthquakes/pager/.
[3] See earthquake.usgs.gov/earthquakes/shakemap/.

USGS
science for a changing world

Earthquake Shaking **Red Alert**

 **USAID**
FROM THE AMERICAN PEOPLE

GSN
ANSS

M 6.3, SOUTH ISLAND OF NEW ZEALAND
Origin Time: Mon 2011-02-21 23:51:43 UTC (12:51:43 local)
Location: 43.60°S 172.71°E Depth: 5 km

PAGER
Version 7
Created: 3 days, 22 hours after earthquake

Estimated Fatalities

Red alert level for economic losses. Extensive damage is probable and the disaster is likely widespread. Estimated economic losses are 10-70% GDP of New Zealand. Past events with this alert level have required a national or international level response.

Yellow alert level for shaking-related fatalities. Some casualties are possible.

Estimated Economic Losses

Estimated Population Exposed to Earthquake Shaking

ESTIMATED POPULATION EXPOSURE (k = x1000)	- -*	23*	46k*	91k	50k	63k	228k	92k	0
ESTIMATED MODIFIED MERCALLI INTENSITY	I	II-III	IV	V	VI	VII	VIII	IX	X+
PERCEIVED SHAKING	Not felt	Weak	Light	Moderate	Strong	Very Strong	Severe	Violent	Extreme
POTENTIAL DAMAGE — Resistant Structures	none	none	none	V. Light	Light	Moderate	Moderate/Heavy	Heavy	V. Heavy
POTENTIAL DAMAGE — Vulnerable Structures	none	none	none	Light	Moderate	Moderate/Heavy	Heavy	V. Heavy	V. Heavy

*Estimated exposure only includes population within the map area.

Population Exposure

population per ~1 sq. km from Landscan

0   5   50   100   500   1000   5000   10000

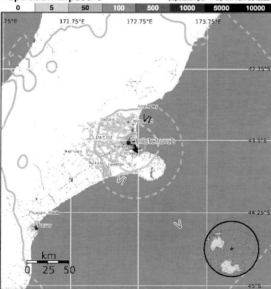

Structures:
Overall, the population in this region resides in structures that are highly resistant to earthquake shaking, though some vulnerable structures exist. The predominant vulnerable building types are reinforced masonry and concrete/cinder block masonry construction.

Historical Earthquakes (with MMI levels):

Date (UTC)	Dist. (km)	Mag.	Max MMI(#)	Shaking Deaths
1994-06-19	90	5.9	VIII(12)	0
1984-06-24	159	6.1	VIII(18)	0
1990-02-10	134	6.0	VIII(61)	0

Selected City Exposure
from GeoNames.org

MMI	City	Population
IX	Christchurch	364k
VII	Lincoln	2k
VI	Woodend	3k
VI	Rolleston	3k
VI	Burnham	1k
VI	Leeston	1k
VI	Oxford	2k
V	Darfield	2k
IV	Timaru	28k
IV	Greymouth	9k
IV	Hokitika	3k

bold cities appear on map   (k = x1000)

PAGER content is automatically generated, and only considers losses due to structural damage.
Limitations of input data, shaking estimates, and loss models may add uncertainty.
http://earthquake.usgs.gov/pager

Event ID: usb0001igm

FIGURE 1.4 Sample PAGER output for the strong and damaging February 2011 earthquake in Christchurch, New Zealand. SOURCE: USGS. Available at earthquake.usgs.gov/earthquakes/pager/events/us/b0001igm/index.html.

oping more precise loss estimation models and related decision support tools (e.g., HAZUS). The vulnerabilities are increasingly documented through state-of-the-art geospatial and temporal methods (e.g., GIS, remote sensing, and visual overlays of hazardous areas with demographic information), and the resulting data are equally relevant to pre-, trans-, and post-disaster social science investigations.

• **Household and business-sector adoption of self-protective measures.** A solid knowledge base has been developed under NEHRP at the household level on vulnerability assessment, risk communication, warning response (e.g., evacuation), and the adoption of other forms of protective action (e.g., emergency food and water supplies, fire extinguishers, procedures and tools to cut off utilities, hazard insurance). Adoption of these and other self-protective measures has been modeled systematically, highlighting the importance of disaster experience and perceptions of personal risk (i.e., beliefs about household vulnerability to and consequences of specific events) and, to a lesser extent, demographic variables (e.g., income, education, home ownership) and social influences (e.g., communications patterns and observations of what other people are doing). Although research on adoption of self-protective measures of businesses is much more limited, recent experience of disaster-related business or lifeline interruptions has been shown to be correlated with greater preparedness activities, at least in the short run. Such preparedness activities are more likely to occur in larger as opposed to smaller commercial enterprises.

• **Public-sector adoption of disaster mitigation measures.** Most NEHRP-sponsored social science research has focused on the politics of hazard mitigation as they relate to intergovernmental issues in land-use regulations. The highly politicized nature of these regulations has been well documented, particularly when multiple layers of government are involved. Governmental conflicts regarding responsibility for the land-use practices of households and businesses are compounded by the involvement of other stakeholders (e.g., bankers, developers, industry associations, professional associations, other community activists, and emergency management practitioners). The results are complex social networks of power relationships that constrain the adoption of hazard mitigation policies and practices at local and regional levels.

• **Hazard insurance issues.** NEHRP-sponsored social research has documented many difficulties in developing and maintaining an actuarially sound insurance program for earthquakes and floods—those who are most likely to purchase earthquake and flood insurance are, in fact, those who are most likely to file claims. This problem makes it virtually impossible to sustain an insurance market in the private sector for these hazards. Economists and psychologists have documented in laboratory studies a number of logical deficiencies in the way people process information related to risks as it relates to insurance decision-making. Market failure in earthquake and flood insurance remains an important social science research and public policy issue.

•   **Public-sector adoption of disaster emergency and recovery preparedness measures.** NEHRP-sponsored social science studies of emergency preparedness have addressed the extent of local support for disaster preparedness, management strategies for improving the effectiveness of community preparedness, the increasing use of computer and communications technologies in disaster planning and training, the structure of community preparedness networks, and the effects of disaster preparedness on both pre-determined (e.g., improved warning response and evacuation behavior) and improvised (e.g., effective ad hoc uses of personnel and resources) responses during actual events. Thus far there has been little social science research on the disaster recovery aspect of preparedness.

•   **Social impacts of disasters.** A solid body of social science research supported by NEHRP has documented the destructive impacts of disasters on residential dwellings and the processes people go through in housing recovery (emergency shelter, temporary sheltering, temporary housing, and permanent housing), as well as analogous impacts on businesses. Documented specifically are the problems faced by low-income households, which tend to be headed disproportionately by females and racial or ethnic minorities. Notably, there has been little social science research under NEHRP on the impacts of disasters on other aspects of the built environment. There is a substantial research literature on the psychological, social, and economic and (to a lesser extent) political impacts of disaster, which suggests that these impacts, while not random within impacted populations, are generally modest and transitory.

•   **Post-disaster responses by the public and private sectors.** Research before and since the establishment of NEHRP in 1977 has contradicted misconceptions that during disasters, panic will be widespread, that large percentages of those who are expected to respond will simply abandon disaster relief roles, that local institutions will break down, that crime and other forms of anti-social behavior will be rampant, and that the mental impairment of victims and first responders will be a major problem. Existing and ongoing research is documenting and modeling the mix of expected and improvised responses by emergency management personnel, the public and private organizations of which they are members, and the multi-organizational networks within which these individual and organizational responses are nested. As a result of this research, a range of decision support tools is now being developed for emergency management practitioners.

•   **Post-disaster reconstruction and recovery by the public and private sectors.** Prior to NEHRP relatively little was known about disaster recovery processes and outcomes at different levels of analysis (e.g., households, neighborhoods, firms, communities, and regions). NEHRP-funded projects have helped to refine general conceptions of disaster recovery, made important contributions in understanding the recovery of households and communities (primarily) and businesses (more recently), and contributed to the development of statistically based community and regional models of post-disaster losses and recovery processes.

•   **Research on resilience has been a major theme of the NSF-supported earthquake research centers.** The Multidisciplinary Center for Earthquake Engineering Research (MCEER) sponsored research providing operational definitions of resilience, measuring its cost and effectiveness, and designing policies to implement it at the level of the individual household, business, government, and nongovernment institution. The Mid-American Earthquake Center (MAE) sponsored research on the promotion of earthquake-resilient regions.

ROADMAP CONTEXT—
THE EERI REPORT AND NEHRP STRATEGIC PLAN

The 2008 NEHRP Strategic Plan calls for an accelerated effort to develop community resilience. The plan defines a vision of "a nation that is earthquake resilient in public safety, economic strength, and national security," and articulates the NEHRP mission "to develop, disseminate, and promote knowledge, tools, and practices for earthquake risk reduction—through coordinated, multidisciplinary, interagency partnerships among NEHRP agencies and their stakeholders—that improve the Nation's earthquake resilience in public safety, economic, strength, and national security." The plan identifies three goals with fourteen objectives (listed below), plus nine strategic priorities (presented in Appendix A).

Goal A: Improve understanding of earthquake processes and impacts.

Objective 1: Advance understanding of earthquake phenomena and generation processes.

Objective 2: Advance understanding of earthquake effects on the built environment.

Objective 3: Advance understanding of the social, behavioral, and economic factors linked to implementing risk reduction and mitigation strategies in the public and private sectors.

Objective 4: Improve post-earthquake information acquisition and management.

Goal B: Develop cost-effective measures to reduce earthquake impacts on individuals, the built environment, and society-at-large.

Objective 5: Assess earthquake hazards for research and practical application.

Objective 6: Develop advanced loss estimation and risk assessment tools.

Objective 7: Develop tools that improve the seismic performance of buildings and other structures.

Objective 8: Develop tools that improve the seismic performance of critical infrastructure.

Goal C: Improve the earthquake resilience of communities nationwide.

Objective 9: Improve the accuracy, timeliness, and content of earthquake information products.

Objective 10: Develop comprehensive earthquake risk scenarios and risk assessments.

Objective 11: Support development of seismic standards and building codes and advocate their adoption and enforcement.

Objective 12: Promote the implementation of earthquake-resilient measures in professional practice and in private and public policies.

Objective 13: Increase public awareness of earthquake hazards and risks.

Objective 14: Develop the nation's human resource base in earthquake safety fields.

Although the Strategic Plan does not specify the activities that would be required to reach its goals, in the initial briefing to the committee NIST, the NEHRP lead agency, described the 2003 report by the EERI, *Securing Society Against Catastrophic Earthquake Losses*, as at least a starting point. The EERI report lists specific activities—and estimates costs—for a range of research programs (presented in Appendix B) that are in broad accord with the goals laid out in the 2008 NEHRP Strategic Plan. The committee was asked to review, update, and validate the programs and cost estimates laid out in the EERI report.

COMMITTEE CHARGE AND SCOPE OF THIS STUDY

The National Institute of Standards and Technology—the lead NEHRP agency—commissioned the National Research Council (NRC) to undertake a study to assess the activities, and their costs, that would be required for the nation to achieve earthquake resilience in 20 years (Box 1.2). The charge

---

BOX 1.2
Statement of Task

A National Research Council committee will develop a roadmap for earthquake hazard and risk reduction in the United States. The committee will frame the road map around the goals and objectives for achieving national earthquake resilience in public safety and economic security stated in the current strategic plan of the National Earthquake Hazard Reduction Program (NEHRP) submitted to Congress in 2008. This roadmap will be based on an analysis of what will be required to realize the strategic plan's major technical goals for earthquake resilience within 20 years. In particular, the committee will:

• Host a national workshop focused on assessing the basic and applied research, seismic monitoring, knowledge transfer, implementation, education, and outreach activities needed to achieve national earthquake resilience over a twenty-year period.
• Estimate program costs, on an annual basis, that will be required to implement the roadmap.
• Describe the future sustained activities, such as earthquake monitoring (both for research and for warning), education, and public outreach, which should continue following the 20-year period.

---

to the committee recognized that there would be a requirement for some sustained activities under the NEHRP program after this 20-year period.

To address the charge, the NRC assembled a committee of 12 experts with disciplinary expertise spanning earthquake and structural engineering; seismology, engineering geology, and earth system science; disaster and emergency management; and the social and economic components of resilience and disaster recovery. Committee biographic information is presented in Appendix C.

The committee held four meetings between May and December, 2009, convening twice in Washington, DC; and also in Irvine, CA; and Chicago, IL (see Appendix D). The major focal point for community input to the committee was a 2-day open workshop held in August 2009, where concurrent breakout sessions interspersed with plenary addresses enabled the committee to gain a thorough understanding of community perspectives regarding program needs and priorities. Additional briefings by NEHRP agency representatives were presented during open sessions at the initial and final committee meetings.

Report Structure

Building on the 2008 NEHRP Strategic Plan and the EERI report, this report analyses the critical issues affecting resilience, identifies challenges and opportunities in achieving that goal, and recommends specific actions that would comprise a roadmap to community resilience. Because the concept of "resilience" is a fundamental tenet of the roadmap for realizing the major technical goals of the NEHRP Strategic Plan, Chapter 2 presents an analysis of the concept of resilience, a description of the characteristics of a resilient community, resilience metrics, and a description of the benefits to the nation of a resilience-based approach to hazard mitigation. Chapter 3 contains descriptions of the 18 broad, integrated tasks comprising the elements of a roadmap to achieve national earthquake resilience focusing on the specific outcomes that could be achieved in a 20-year timeframe, and the elements realizable within 5 years. These tasks are described in terms of the proposed activity and actions, existing knowledge and current capabilities, enabling requirements, and implementation issues. Costs to implement these 18 tasks are presented in Chapter 4, in as much detail as possible within the constraint that some components have been the subject of specific, detailed costing exercises whereas others are necessarily broad-brush estimates at this stage. The final chapter briefly summarizes the major elements of the roadmap.

2

What Is National Earthquake Resilience?

The concept of "resilience" is fundamental to a roadmap for realizing the major technical goals of the 2008 NEHRP Strategic Plan within 20 years. The Strategic Plan articulates a vision, mission, and goals that aim to "improve the nation's earthquake-resilience in public safety, economic strength, and national security" (NIST, 2008; p. iii). The meaning of "resilience," however, is far from clear. Numerous definitions of "resilience" exist, and the term is often used loosely and inconsistently. To provide a context and vision for the roadmap, this chapter sets out a working definition of "national earthquake resilience" that includes a brief discussion of conceptual and measurement issues. The discussion draws on committee discussions, the rapidly expanding literature on resilience, and input from more than 50 leading earthquake professionals at an August 2009 workshop sponsored by the committee. Two examples are then provided—Evansville, Indiana, and San Francisco, California—to illustrate how a community might work toward a vision of resilience.

DEFINING NATIONAL EARTHQUAKE RESILIENCE

Dozens of definitions of "resilience" can now be found in the literature, reflecting a range of perspectives and a lack of consensus on the meaning of the term. In the context of hazards and disasters, three definitions of resilience that are often cited are:

> The capability of an asset, system, or network to maintain its function or recover from a terrorist attack or any other incident (DHS, 2006).

The capacity of a system, community or society potentially exposed to
hazards to adapt, by resisting or changing in order to reach and maintain
an acceptable level of functioning and structure. This is determined by
the degree to which the social system is capable of organizing itself to
increase this capacity for learning from past disasters for better future
protection and to improve risk reduction measures (UN ISDR, 2006; also
SDR, 2005).

The ability of social units (e.g., organizations, communities) to mitigate
risk and contain the effects of disasters, and carry out recovery activities
in ways that minimize social disruption while also minimizing the effects
of future disasters. Disaster Resilience may be characterized by reduced
likelihood of damage to and failure of critical infrastructure, systems,
and components; reduced injuries, lives lost, damage, and negative economic and social impacts; and reduced time required to restore a specific
system or set of systems to normal or pre-disaster levels of functionality
(MCEER, 2008).

Of these, the Department of Homeland Security's National Infrastructure Protection Program (NIPP) definition is narrower in scope than the
MCEER (Multidisciplinary Center for Earthquake Engineering Research)
definition, and the concept of maintaining function is somewhat vague in
the former. It could include maintaining as high a function as possible at
the moment the disaster strikes. Alternatively, resilience might refer only
to maintaining function through activities undertaken after the event,
and hence would not necessarily include pre-event mitigation. This focus
on post-shock activities (*both inherent and adaptive*) and the emphasis on
recovery as both goal and process are more consistent with the origins
of the term resilience. The United Nations International Strategy for
Disaster Reduction (ISDR) definition, in contrast, departs further from
the origins of the term and appears to emphasize pre-disaster mitigation
and preparedness, with the only allusion to the idea of rebounding from a
disaster relating to the speed of recovery. It does, however, emphasize that
resilience is a process. This definition is also used in the National Science
and Technology Council's *Grand Challenges for Disaster Reduction*.
    Although the 2008 NEHRP Strategic Plan (NIST, 2008; p.47) adopts
this latter definition, for purposes of the roadmap, it is important to consider several issues:

  • "National earthquake resilience" should primarily involve building resilience at the level of *communities*. It is also important, however, to
prepare for the rare instances where earthquake disasters could extend
beyond localities and have national-level consequences (see Box 2.1).
  • In order for communities to be more resilient, support from both
state and federal levels is required.

- Building national earthquake resilience should foster synergies between resilience to earthquakes and to other hazards.
- Communities should consider developing multi-tier resilience goals and strategies, i.e., different performance expectations for different scale events. In some cases, it may be effective to focus actions on containing the effects of "expected" events, rather than very rare, "extreme" events.
- Resilience involves both pre-disaster mitigation (activities to reduce the amount of loss in an event) and the ability to mute post-event losses and rapidly recover from an event.
- Resilience should allow for systemic change, especially in low-probability, high-consequence events. Resilience does not necessarily entail a return to "normal" or "pre-disaster" conditions. Reducing future risk should also be a goal of recovery activities.

With these considerations in mind, the committee recommends that NEHRP adopt the following working definition for "national earthquake resilience" (applicable more generally to all-hazards resilience):

A disaster-resilient nation is one in which its communities, through mitigation and pre-disaster preparation, develop the adaptive capacity to maintain important community functions and recover quickly when major disasters occur.

MEASURING DISASTER RESILIENCE

Reflecting the lack of a consensus definition, no standard metric exists for measuring disaster resilience. Indeed, one of the priorities in the National Science and Technology Council's (NSTC's) *Grand Challenges for Disaster Reduction* is to "assess disaster resilience using standard methods" (SDR, 2005; p. 2). As this report noted, such metrics are needed for several reasons: "With consistent factors and regularly updated metrics, communities will be able to maintain report cards that accurately assess the community's level of disaster resilience. This, in turn, will support comparability among communities and provide a context for action to further reduce vulnerability. Validated models, standards, and metrics are needed for estimating cumulative losses, projecting the impact of changes in technology and policies, and monitoring the overall estimated economic loss avoidance of planned actions" (SDR, 2005; p. 2). Perhaps most importantly, standardized methods are needed to gauge improvements in resilience as a result of disaster risk reduction planning and mitigation.

Metrics of disaster resilience differ from the familiar metrics of disaster risk in several ways. Standard risk measures include expected casualties, property damage, and business interruption loss—that is, estimates of

BOX 2.1
Widespread Consequences of a Central U.S. Earthquake

An analysis of the impacts of a magnitude-7.7 earthquake on all three New Madrid faults was performed by the Mid-America Earthquake Center under the FEMA New Madrid Catastrophic Planning Initiative (Elnashai et al., 2009). Results indicated that this event would have widespread, catastrophic consequences (Figure 2.1), including:

- Nearly 715,000 buildings damaged in eight states.
- Substantial damage to critical infrastructure (essential facilities, transportation, and utility lifelines) in 140 counties: 2.6 million households without electric power; 425,000 breaks and leaks to both local and interstate pipelines; and 3,500 damaged bridges, with 15 major bridges unusable.
- 86,000 casualties for a 2:00 am scenario, with 3,500 fatalities.
- 7.2 million people displaced, with 2 million seeking temporary shelter.
- 130 hospitals damaged.
- $300 billion in direct economic losses, including buildings, transportation, and utility lifelines, but excluding business interruption costs.

Moreover, infrastructure damage would have a major impact on interstate transport crossing the Central United States.

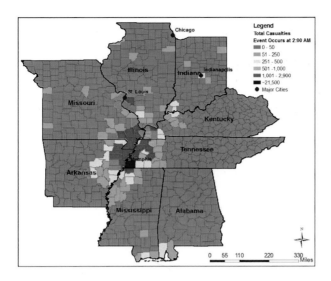

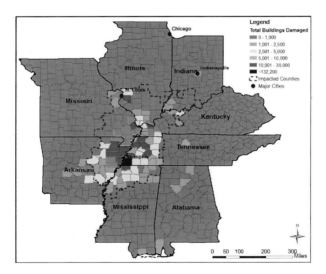

FIGURE 2.1 Distribution of top) the nearly 86,000 total casualties, including 3,500 fatalities, and bottom) the more than 713,000 buildings damaged, in the eight-state study region from a magnitude-7.7 scenario earthquake at 2:00 am on the New Madrid faults. SOURCE: Elnashai et al. (2009); Courtesy of the Mid-America Earthquake Center, University of Illinois.

these losses in potential earthquakes weighted by the probability of such events occurring. Resilience differs from risk in three important ways. First, resilience includes performance in the post-disaster (response and recovery) timeframes, including aspects such as business interruption and the time required to recover, while risk typically focuses on immediate property damage. Second, resilience embodies some sense of goals and considerations of what risk is acceptable. Third, it also encompasses ideas of capacity-building and process, rather than being limited in scope to goals and outcomes.

Because the concept of resilience is specific to the context of the specific community and its goals, it can be expected that no single measure will be able to capture it sufficiently. Moreover, different measures will be needed for different purposes. Thus for federal agencies, a national-scale overview may be useful; a simple measure might be the percentage of these states with active seismic safety programs. For a state government, a useful marker may be the percentage of communities that are actively engaging in seismic risk reduction. For a city, however, more specific measures would be needed. An overall metric of the time required to recover "community wellness" (e.g., an aggregation of casualties, property, and economic losses) in the event of an "expected" earthquake may be one possibility. Annualized expected earthquake losses in that community may provide another alternative. Within a community, organizations such as local fire departments may have yet more specific measures in relation to seismic performance goals. Thus multi-level assessments are needed, rather than searching for a "one size fits all" metric.

Researchers and practitioners have proposed a number of approaches for measuring disaster resilience at the community level. These approaches can be broadly categorized into two types—those emphasizing resilience as a goal, and those emphasizing it as a process. A few examples are briefly reviewed here.

Bruneau et al. (2003), on which the NEHRP definition of resilience is based, treats resilience in terms of performance outcomes or goals. They propose as a measure of resilience the functional or performance loss of a system (such as a city) evaluated over the timeframe for recovery. This is illustrated schematically in Figure 2.2. The smaller the initial drop in a disaster, and the more rapid the recovery, the smaller the aggregate loss ("loss triangle") and the higher the assessed resilience.

Within this framework, recovery is assumed to entail a return to normal (without-disaster) conditions. Thus, it is difficult to address some of the aspects of resilience discussed above, such as allowing for system change and rebuilding in ways that reduce future risk. However, the framework can be generalized to accommodate these considerations. A summary of recent progress includes:

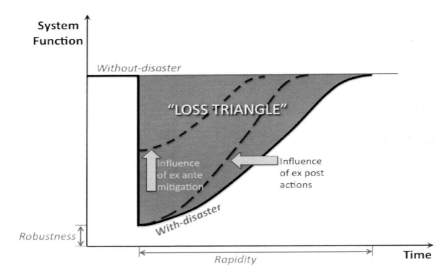

FIGURE 2.2 Measuring resilience using the "Loss Triangle" concept. Note that the degree of "robustness" depends upon both the system's inherent resilience and the additional effect of any pre-disaster mitigation actions. SOURCE: Modified from Bruneau et al. (2003) and McDaniels et al. (2008). Reprinted from McDaniels et al. (2008) with permission from Elsevier.

- Several researchers have proposed operational metrics (e.g., Chang and Shinozuka, 2004; Rose, 2004, 2007). The most basic of these provides a starting point for measurement as the avoided losses due to resilience actions divided by the maximum potential losses for a given event.
- An important distinction has been made between system resilience and broader concepts such as economic resilience. The latter is more encompassing because it focuses on the contribution these services make to the economy, including not just the supply but also demand (not just to the first line of customers but also to successive ones down the customer chain, e.g., Cox et al. (2011).
- Recent programs have embraced the resilience concept. The SPUR (San Francisco Planning and Urban Research Association–Resilient City Initiative) approach (SPUR, 2009) also focuses on outcomes (see Figure 2.3 and related discussion). Data for these outcomes are derived, however, from expert judgments, rather than either community consultation or a computer model.
- Broader measures of resilience emphasize the capacity, or process, dimensions of resilience. These typically characterize resilience through

TARGET STATES OF RECOVERY FOR SAN FRANCISCO'S BUILDING AND INFRASTRUCTURE

INFRASTRUCTURE CLUSTER FACILITIES	Event Occurs	Phase 1 Hours			Phase 2 Days		Phase 3 Months		
		4	24	72	30	60	4	36	36+
CRITICAL RESPONSE FACILITIES AND SUPPORT SYSTEMS									
Hospitals	▨							X	
Police and fire stations	▨		X						
Emergency operations center	X								
Related utilities		▨				X			
Roads and ports for emergency		▨		X					
CalTrain for emergency traffic			▨		X				
Airport for emergency traffic				X					
EMERGENCY HOUSING AND SUPPORT SYSTEMS									
95% residence shelter-in-place		▨						X	
Emergency Responder Housing		▨		X					
Public shelters		▨					X		
90% Related Utilities			▨					X	
90% roads, port facilities, and public transit			▨				X		
90% Muni and BART Capacity			▨			X			
HOUSING AND NEIGHBORHOOD INFRASTRUCTURE									
Essential city service facilities				▨			X		
Schools					▨		X		
Medical provider offices								X	
90% neighborhood retail services									X
95% of all utilities								X	
90% roads and highways						X			
90% transit						X			
90% railroads								X	
Airport for commercial traffic					X				
95% transit						▨	X		
COMMUNITY RECOVERY									
All residences repaired, replaced or relocated							▨	X	
95% neighborhood retail businesses open								X	
50% offices and workplaces open									X
Non-emergency city service facilities							▨		
All businesses open									X
100% utilities									X
100% highway and roads									X
100% transit									X

Source: SPUR Urbanist, February 2009

The "x's" in the chart to the right indicate SPUR's best educated guesses about current standards for recovery times. The shaded areas represent the goals — targets based on clearly stated performance measures (see next page) — for recovery times for the city's buildings and lifelines. The gaps between "x's" and shaded boxes represent how far we are from meeting resiliency targets.

TARGET STATES OF RECOVERY

Performance Measure	Description of usability after expected event	
	BUILDINGS	LIFELINES
(Category A)	Category A: Safe and operational	
(Category B)	Category B: Safe and usable during repairs	100% restored in 4 hours
(Category C)	Category C: Safe and usable after moderate repairs	100% restored in 4 months
(Category D)	Category D: Safe and usable after major repairs	100% restored in 3 years
X	Expected current status	

FIGURE 2.3 Resilience goals in San Francisco described in the policy paper adopted by the Board of the San Francisco Planning and Urban Research Association (SPUR, 2009). SOURCE: SPUR (2009).

describing features of more disaster-resilient communities or identifying specific actions, adaptations, or tactics both pre- and post-disaster (e.g., Tobin, 1999; Godschalk, 2003; Berke and Campanella, 2006; Cutter et al., 2008a, 2008b; Norris et al., 2008). More recently, progress has been made on developing indices of community resilience (e.g., Emmer, 2008; Cutter et al., 2010; CARRI, 2011). These prospective measures of resilience are facilitated by the use of census or other generally available data and self assessments.

These examples illustrate the range of approaches that have been applied to assess the disaster resilience of communities. As noted earlier, no one resilience indicator can suit all purposes, and different measurement approaches may be appropriate in different contexts for assessing current levels of disaster resilience and incremental progress in developing resilience.

WHAT DOES AN EARTHQUAKE-RESILIENT COMMUNITY LOOK LIKE?

The NSTC's *Grand Challenges for Disaster Reduction* identified four key characteristics of disaster-resilient communities (SDR, 2005; p. 1):[1]

- Relevant hazards are recognized and understood.
- Communities at risk know when a hazard event is imminent.
- Individuals at risk are safe from hazards in their homes and places of work.
- Disaster-resilient communities experience minimum disruption to life and economy after a hazard event has passed.

Within the context of this broad vision, more specific, tangible characterizations of a more earthquake-resilient community are proposed here in order to guide prioritization of efforts. In a major disaster:

- **No systematic concentration of casualties.** Important or high-occupancy structures (e.g., schools, hospitals, and other major institutional buildings; high-rise commercial and residential buildings) do not collapse, and significant numbers of specific building types (e.g., hazardous unreinforced masonry structures) do not collapse. There are no major hazardous materials releases that would cause mass casualties.

---

[1] A number of other similar characterizations have also been proposed (e.g., Godschalk, 2003; Foster, 2007). Tierney (workshop presentation) notes that resilience has multiple aims—reduced loss of life and economic impact; equity and fairness (addressing disparities in vulnerability); and sustainability (laws, processes, etc. are robust over time and support social values of quality of life, environmental quality, community safety, and livability).

- **Financial loss and societal consequences are manageable, not catastrophic.** Damage to the built environment is reduced to avoid catastrophic financial and societal losses due to overwhelming cost of repair, casualties, displaced populations, government interruption, loss of housing, or loss of jobs. Community character and cultural values are maintained following disasters; there is not wholesale loss of iconic buildings (including those designated as historic), groups of buildings, and neighborhoods of architectural, historic, ethnic, or other significance.

- **Emergency responders are able to respond and improvise.** Roads are passable, fire suppression systems are functional, hospitals and other critical facilities are functional. It is noteworthy that during the 9/11 attacks, New York City's response was hampered by the need to set up a new Emergency Operations Center, the existing one having been located in the World Trade Center.

- **Critical infrastructure services continue to be provided in the aftermath of a disaster.** Energy, water, and transportation are especially critical elements. Telecommunications are also very important. Continued service is needed for critical facilities such as hospitals to function, as well as for households to remain sheltered in their homes.

- **Disasters do not escalate into catastrophes.** Infrastructure interdependencies have been anticipated and mitigated, so that disruptions to one critical infrastructure do not cause cascading failures in other infrastructures (e.g., levee failures in New Orleans escalated the disaster into a catastrophe). Fires are quickly contained and do not develop into major urban conflagrations that cause mass casualties and large-scale neighborhood destruction.

- **Resources for recovery meet the needs of all affected community members.** Resources for recovery are available in an adequate, timely, and equitable manner. To a large extent, local governments, nonprofit organizations, businesses, and residents would have already materially and financially prepared for a major disaster (e.g., are adequately insured; have undertaken resilience activities on their own and in cooperation with others). Safety nets are in place for the most vulnerable members of society.

- **Communities are restored in a manner that makes them more resilient to the next event.** Experience is translated into improved design, preparedness, and overall resilience. High-hazard areas are rebuilt in ways that reduce, rather than recreate, conditions of disaster vulnerability.

Each community will face unique gaps and challenges in meeting these resilience goals. The priorities and mix of strategies and actions will differ from one community to the next. Each community could translate these general goals into specific, transparent performance goals appropriate for the locality and scaled for different size disasters. These performance goals can then provide a basis for developing consistent design standards and retrofit guidelines.

Two examples are provided below to illustrate different approaches that proactive communities have undertaken to enhance their disaster resilience. The Evansville, Indiana, example is noteworthy for the long-term, cumulative efforts of multiple stakeholder groups. Evansville focused largely on traditional pre-disaster mitigation and planning actions—that is, enhancing "robustness" as noted in Figure 2.2. In contrast, the San Francisco example is noteworthy for pioneering community discussions and prioritizing activities that focus explicitly on the "rapidity" dimension of resilience in the aftermath of an earthquake.

Example 1: The Process of Developing Resilience in Evansville

This example outlines the history of Evansville, Indiana's Disaster Resistant Community (DRC) efforts as an example of one community's long-term, multi-faceted approach to developing disaster resilience. After a 1987 central U.S. earthquake and the 1989 Loma Prieta earthquake, geologists and emergency response planners recognized that Evansville, Indiana, was at greater risk from earthquakes than most Indiana cities because parts of the city are built upon thick soft soils. In 1990, long before the national programs to improve resiliency of communities, Evansville started its own effort with support by the Indiana Department of Fire and Building Services (IDFBS) and the City of Evansville. Initial activities involved gathering subsurface soil property information by the Indiana Geological Survey and Ball State University. The geologic, geotechnical, and shear wave velocity data provided the basis for risk analysis for the IDFBS and Vanderburgh County Building Commission and emergency management response planning.

In 1997, the Central U.S. Earthquake Consortium (CUSEC) embarked on a pilot disaster-resistant community project involving two communities— Evansville, Indiana, and Henderson, Kentucky. To launch the pilot project, a workshop was held to bring together a multi-disciplinary group of hazards specialists, emergency managers, and community leaders to develop a model disaster-resistant community program. This workshop was co-sponsored by Federal Emergency Management Administation (FEMA), Insurance Institute for Property Loss Reduction (IIPLR), and the Disaster Recovery Business Alliance along with the cooperating organizations of the American Red Cross, Risk Management Solutions, Inc., International City and County Management Association, and Evansville community leaders. Working groups developed a mitigation strategy and implementation plan that addressed the key elements of a DRC program: Education and Public Outreach, Existing Development, New Development, Com-

munity Land Use, and Business Vulnerability Reduction. A steering committee identified three key components of an Evansville Model Disaster
Resistant Community Program: (1) use of the HAZUS loss estimation
software as a central feature of the community's hazard and risk assessment; (2) application to become a "Showcase Community" in a national
program administered by the IIPLR; and (3) formation of an Evansville
Business Alliance. The committee outlined objectives and sample activities
with the recognition that becoming more disaster resistant would require
a long-term, phased approach under the guidance of a partnership of local
and national interests.

In applying for the Showcase Community program, the committee
agreed to meet 14 criteria:

1. Adopt the latest model building code without modifications.
2. Receive the Building Code Effectiveness Grading Schedule grade
and develop an improvement strategy.
3. Participate in the National Flood Insurance Program, and receive a
Community Rating Service grade and develop an improvement strategy.
4. Have a minimum of 8 on the fire suppression rating system.
5. Undergo a community risk assessment conducted by the IIPLR
and the partnership of local and national interests.
6. Develop and offer mitigation training to professionals (e.g., engineers, architects, building officials, contractors).
7. Conduct nonstructural retrofit assessment of all nonprofit child
care centers so that the partnership can retrofit them.
8. Provide public education of natural hazards and mitigation techniques to certify homeowners to qualify them for incentives.
9. Develop K-12 school curriculum teaching about natural hazard
risks and mitigation.
10. Ensure that the community has a land-use plan and a planner, and
makes zoning decisions in compliance with its land-use plan.
11. Develop an emergency recovery plan and post-disaster recovery
plan.
12. Develop a Disaster Recovery Business Alliance to formulate and
implement a business mitigation strategy.
13. Develop public- and private-sector incentives.
14. Participate in the Partnership Seal of Approval inspection and
certification.

The steering committee worked with the Institute for Business and
Home Safety (IBHS) to complete the list of projects. Upon completion in
1997, Evansville was named the nation's first Showcase Community.

In 1998, Evansville applied to FEMA to be part of Project Impact and was chosen in the second round. When the grant from FEMA was received, a decision was made to incorporate into the Southwest Indiana Disaster Resistant Community Corporation (DRC). The nonprofit corporation has representation from five counties in southwestern Indiana. In the fall of 1997, a movement to develop an alliance of area businesses was begun by the executive director of the Metropolitan Evansville Chamber of Commerce and other regional business executives. The Southwest Indiana Disaster Recovery Business Alliance (DRBA) was to develop disaster recovery initiatives. This effort was a good fit with the DRC, and a combined office was established with a full-time director in 1999.[2]

The DRC efforts resulted in numerous accomplishments, a few examples of which are highlighted below to illustrate the range of partnerships involved, the types of activities undertaken, and the spillover benefits of earthquake risk reduction to multi-hazard resilience activities:

- Seismic retrofits were completed in critical and other facilities. Several fire stations were structurally and nonstructurally retrofitted. Nonstructural retrofits were completed at 36 nonprofit daycare centers using materials donated by area businesses and labor provided by volunteers from the local building commission, a youth group, insurance agencies, and the DRC. The school corporation adopted several mitigation policies and was involved in building the ECO House, the first house to be certified "disaster resistant" by the Institute for Business and Home Safety. The DRC coordinated volunteers in nonstructural mitigation of dozens of Habitat for Humanity homes. The City of Evansville's Housing Rehab Services agreed to strap down all water heaters as part of its housing rehabilitation program provided to low-moderate income households.
- Other accomplishments served to incorporate hazard and risk considerations into urban development. The Area Plan Commission considered hazard and loss estimation information in updating the comprehensive plan. Evansville-Vanderburgh County committed to taking natural hazards into account in all its land-use decisions. A new building code amendment required new buildings to be constructed to withstand 110 mph winds. Vanderburgh County and Evansville, Indiana, received National Flood Insurance Program community ratings that provided 10%

---

[2] The uniqueness of the Evansville DRC gained other national recognition. For example, Sandia National Laboratories partnered with the DRC in 1999 to develop a disaster management system proposal to identify vulnerability issues related to critical infrastructures. Also, the U.S. Geological Survey picked Evansville in 2003 for one of its Urban Hazard Mapping Programs. The project's goal is to provide state-of-the-art urban seismic hazard maps reflecting the variations in materials and thicknesses that govern the amount of amplification by the soils and locations of liquefaction. The scenario earthquakes, representing reasonable maximum magnitude earthquakes for these areas, are being used to produce ground motion and liquefaction potential maps.

and 5% reductions, respectively, in National Flood Insurance premiums for local residents.

•   Training sessions were conducted for professionals. These included city-county building officials, architects and engineers, and fire department personnel. The HAZUS initiative involved numerous participants. Data development involved University of Evansville students, the Indiana Geological Survey, and the Disaster Recovery Business Alliance, among others. Training workshops and a HAZUS Technical Subcommittee were formed to develop and maintain the capacity to use HAZUS for hazard and risk assessment.

•   The DRC and its partners developed and disseminated disaster preparedness and mitigation information to educate the general public. Print materials included a disaster preparedness calendar and mitigation tip sheets by the Southern Indiana Gas & Electric Company and the Red Cross. Fox 7 produced a documentary of the Project Impact initiative. The DRC worked with local schools to incorporate K-12 educational programs on disaster preparedness, response, and mitigation.

•   Members of the DRC organized a number of community events— including Earthquake Preparedness Week, Fire Prevention Week, Severe Weather Week, and Building Safety Week—and participated in others, such as CPR/Family Safety Day and a local hospital's safety fair. These events provided opportunities to educate local residents on preparedness and mitigation.

Even at the end of 2009, with no funding, DRC participants continued to perform walk-through inspections in schools and businesses for preparedness, as well as make presentations to various groups and have a presence at area fairs.

The Evansville plan is admirable for its attention to major concerns of reducing the losses from earthquakes and for moving toward an all-hazards approach. However, it focuses almost entirely on pre-event mitigation, and only three of its major tenets refer to post-disaster recovery and reconstruction. The emphasis in theory and practice since the time of the development of the Evansville plan has been much more focused on post-disaster resilience as defined in this report—an emphasis on maintaining function of the economy and broader society, as well as hastening recovery. The San Francisco example described below is more in accord with the concepts of resilience described in this report, with its design of pre-disaster mitigation activities—utilizing a broad definition of "performance"—which emphasizes not just a reduction in building damage but also an emphasis on maintaining and restoring the services that buildings provide.

Example 2: Defining Resilience Goals and Measures in San Francisco

In 2006, as part of the activities surrounding the 100-year anniversary of the 1906 Great San Francisco Earthquake, the Earthquake Engineering Research Institute (EERI), Seismological Society of America (SSA), California Emergency Management Agency (CalEMA), and U.S. Geological Survey (USGS) commissioned the development of a comprehensive simulation and analysis of potential losses if a repeat of the 1906 earthquake were to happen now. The report, *When the Big One Strikes Again* (Kircher et al., 2006), estimated that many of Northern California's nearly 10 million residents would be affected. It would cost $90-$120 billion to repair or replace the more than 90,000 damaged buildings and their contents, and as many as 10,000 commercial buildings would sustain major structural damage. Between 160,000 and 250,000 households would be displaced from damaged residences. Depending upon whether the earthquake occurs during the day or night, building collapses would cause 800 to 3,400 deaths, and a conflagration similar in scale to the 1906 fire is possible and could cause an immense loss. Damage to utilities and transportation systems would increase losses by an additional 5% to 15%, and economic disruption from prolonged lifeline outages and loss of functional workspace would cost several times this amount. Considering all loss components, the total price tag for a repeat of the 1906 earthquake is likely to exceed $150 billion. In such a scenario, the city of San Francisco might not be able to recover from the cascading consequences and might lose its central place in the region.

Motivated to reverse this prognosis, earthquake professionals and policy-makers in San Francisco joined forces soon after the conference and began a two-year effort to prioritize policies and actions to help ensure that San Francisco could rebound quickly from a major event. Their efforts resulted in four major policy papers, summarized in "The Resilient City," a policy paper adopted by the Board of the San Francisco Planning and Urban Research Association in 2008 (SPUR, 2009). The panel of experts took a community-wide perspective, describing their vision of resilience as:

> Resilient communities have an ability to govern after a disaster strikes. These communities adhere to building standards that allow the power, water and communications networks to begin operating again shortly after a disaster and that allow people to stay in their homes, travel to where they need to be, and resume a fairly normal living routine within weeks. They are able to return to a "new" normal within a few years . . . (and the disaster) does not become a catastrophe that defies recovery (SPUR, 2009; p. 1).

Key elements of this vision include:

•   Establishment of performance objectives for buildings and lifeline infrastructure systems, including power, gas, water, communications, and transportation.
•   Seismic retrofit of a sufficiently large number of homes so that the vast majority of city residents are able to shelter in place (i.e., remain at home) following an earthquake.
•   Establishment of a Lifelines Council with influence over the preparation of critical services. This council would ensure that the utility services are restored within days of the earthquake.
•   Establishment of a new voluntary rating system, designating Seismic Silver and Seismic Gold buildings, which performs so well that these standards quickly becomes a model for all new housing in the region.
•   Ability of the entire city to get back on its feet in four months.

To achieve this vision, the panel established performance targets for new and existing buildings and lifelines, at different phases in the recovery process, for an "expected" earthquake (ATC, 2010). The panel chose to analyze an "expected" earthquake, rather than an "extreme" event, in order to focus on a large event that can reasonably be expected to occur during the useful life of a structure or lifeline system. It chose a scenario earthquake that was also being used by another seismic study under way in the city, with the expected earthquake being a magnitude-7.2 earthquake on the Peninsula segment of the San Andreas Fault. It also established a series of transparent performance measures, based upon usability, for both buildings and infrastructure after the expected event. For buildings, there are three categories: safe and operational, safe and usable during repairs, and safe and usable after moderate repairs. Relying on expert input, the panel assessed the current status of expected performance of buildings and infrastructure. It then set performance targets for four post-earthquake time periods—immediately, 1 to 7 days, 7 days to 2 months, and 2 to 36 months.

SPUR developed a series of near- and long-term recommendations for existing and new buildings as well as infrastructure by considering: (1) the goals for seismic resilience for each component of the city; (2) the gap between current seismic performance and the goal; and (3) the general level of cost to make the necessary improvements or retrofits. In all cases, SPUR's performance targets require a substantial improvement in seismic performance compared to the current situation. However, SPUR did not recommend that all buildings and infrastructure be upgraded to a level that would make them "damage-proof," as this was assessed to be cost-prohibitive. Instead, by defining an acceptable level of damage for the expected earthquake, it focused its recommendations on those improvements considered most likely to yield a quick recovery or level of resilience desired for each phase of recovery. Recommendations were guided by the recognition that two "missing pieces" needed to be addressed in dealing with the earthquake problem—lifelines (critical infrastructure) and the workforce.

The panel emphasized pre-disaster mitigation actions in its recommendations, but some post-disaster actions would also be required to achieve these performance targets. For example, ensuring that "95% of all residences are deemed to be safe for occupancy within 36 hours after the expected earthquake" would require that enough existing structures be seismically retrofitted so that the vast majority of San Francisco residents would be able to shelter in place. It also required substantial changes to inspection procedures and post-earthquake occupancy standards, because residents would need to be allowed to remain in superficially damaged buildings even if utility services are not functioning.

Earthquakes other than the "expected" one are possible, of course, but, in smaller earthquakes, better performance is expected. In larger, more extreme events, lesser performance will have to be tolerated.

Figure 2.3 provides an example of specific resilience goals recommended by SPUR in San Francisco. The figure indicates the expected performance of buildings and infrastructure if the earthquake were to occur today (marked as X's), the post-earthquake performance targets for each category (shaded boxes), and the gap between them. For example, critical response facilities, such as hospitals, police and fire stations, and emergency operations centers, are categorized as buildings that must be "safe and operational" immediately after the expected earthquake. Currently, these buildings are more likely to be "safe and operational" within 24 hours or, as long as 36 months, after an expected earthquake. For residential housing, buildings must be "safe and usable during repairs" and there is a target to have 95% of residents able to shelter-in-place within 24 hours after an expected earthquake. Currently, it is more likely to take up to 36 months before 95% of San Francisco's residents would be able to re-inhabit their homes after an expected earthquake.

Other Examples of Resilience

The Evansville and San Francisco examples described above both represent concerted public programs to improve earthquake resilience. Such programs are needed because there is a lack of information and awareness of the earthquake threat, and a lack of adequate incentives to address it, when the rewards for the entity undertaking the investment in resilience involve spillover effects to other segments of society. In the latter case of a "public good," the entity making the expenditure cannot capture all of the broader gains, and hence an under-investment occurs from the standpoint of society. Otherwise, in a predominantly market economy like that of the United States, many individual decision-makers and public institutions do make appropriate decisions regarding resilience in response to market signals—the marketplace is an important resource for developing resilience.

Prices reflect the value of economic resources, and price increases following a disaster are often characterized as gouging. Nevertheless, some price increases are warranted and serve as indicators of the increased scarcity of specific goods and services. When markets are working effectively, these price signals need to be considered in making decisions regarding the allocation of resources. When markets are not working effectively, as when market institutions are destroyed or prone to various types of market failure (including price gouging due to asymmetric information or market power), it may be necessary for authorities to override market signals and make decisions with other approaches, such as rationing. This may be the case especially where equity, or fairness, is concerned. Free markets are known to lead to the efficient allocation of resources, but are effectively blind to equity concerns.

Individual decision-makers also capitalize on many types of resilience embodied in the economic system, referred to as "inherent" sources of resilience, including the marketplace itself (NRC, 2007; Rose, 2009). These conditions include inventories of critical materials, the ability to substitute other inputs for those in short supply (e.g., use of bottled or trucked water for piped water serves), and excess productive capacity to be accessed when facilities in use are damaged (e.g., relocating to empty office space or factories). Although many of these types of resilience are taken for granted because they are in place during the normal course of doing business, there is still potential for enhancing them. They also have an advantage in reducing losses over mitigation because they can be accessed at little or no extra cost.

Another category of resilience refers to the ingenuity, or "adaptive ability," that often is inspired by necessity after an earthquake to keep households, business, and government organizations going (e.g., Comfort, 1999; Mileti, 1999). Examples include making organizations more efficient, finding new substitutes for critical materials, and establishing new social networks. They are also part of the nation's resilience capability. They may not require large-scale programs as in the previous case study examples, but they do merit attention and further nurturing. Not all decision-makers are aware of these opportunities, and more generic programs, rather than region-specific ones may be the preferable vehicle. For both inherent and adaptive resilience, the dissemination of information on best practice methods has the potential to be a valuable national project to promote resilience.

A new "business continuity industry" has arisen over the past 10 years, consisting of private-sector professionals that help businesses prepare, clean up, and recover from disasters (the majority of examples relate to information technology backup and business relocation). Such services are especially important to small business, which cannot take advantage of economies of scale or otherwise afford their own in-house hazard professionals.

Another reason for focusing on the role of the individual business or household is the importance that self-reliance can provide. It helps reduce dependence on government bailouts. Flynn (2008) has taken a profound view of this by focusing on how resilience can be "empowering" to the general citizenry.

This discussion of economic considerations and reliance practices is related to the object of resilience—what types of losses are we really trying to reduce. The focus of much of this report is on property damage. However, property damage from earthquakes and most other natural disasters takes place at a given point or short period in time. It is, rather, the flow of goods and services from the property (capital assets) that sustain people's lives. This reduction in the flow of goods and services (often referred to as "business interruption," or BI) starts at the point of the earthquake but continues until recovery is complete. Resilience cannot do anything to reduce the property damage after the event, but can reduce the BI by using remaining resources as effectively as possible and recovering as quickly as possible. When economists and policy-makers talk about indicators of societal well-being, they focus on flow indices such as the BI, which in the grander sense is really just a lay term for a decrease in gross national/ regional product.

Also, resilience can be defined narrowly or holistically. System resilience is usually a good example of the former because it focuses on the maintenance of the service flow. Economic resilience is more encompassing because it focuses on the contribution these services make to the economy. It includes not just the supply but also demand (i.e., both the provision of a good or service and its utilization, and not just to the first line of customers but to successive ones down the customer chain). An example of this dichotomy would be transportation resilience in the aftermath of a natural disaster or terrorist attack. It could begin with consideration of resilient actions by providers of transportation services and then proceed to the resilience of its customers through the alternative modes, telecommuting, and greater reliance on existing inventories (as opposed to new shipments). In the latter case, it is not only the number of trips that is important but also the contribution they make to transportation customers' production levels or well-being. This way, telecommuting, would be viewed as a resilient strategy, because it maintains production (reduces BI) even with fewer trips; otherwise, its contribution might be overlooked (e.g., Cox et al., 2011).

In a similar vein, a recent study of the resilience of the New York City Metropolitan Area economy in the aftermath of 9/11 found its resilience to be very high—72% according to one if the definitions noted in the previous section, because 95% of the 1,100 firms located in the World Trade Center area were able to move to other locations, primarily in the metropolitan area (72% is lower than 95% because of the lost production caused by delays in relocation) (Rose et al., 2009). Thus, temporary locations, often becoming permanent, saved more than $40 billion of gross regional product. To use all of society's resources effectively, such flexibility to use excess building stock (if available) before reconstruction could take place needs to be factored into programs such as the San Francisco example.

Resilience and Post-Earthquake Recovery

The Evansville and SPUR examples described above focused on aspects of the built environment and on advance planning for recovery, but they do not illustrate actions that can be taken after the event to promote resilience in terms of maintaining function of the broad set of societal attributes and hastening recovery. Table 2.1 provides examples of resilient actions at various stages of recovery and reconstruction in relation to a broader set of societal attributes and indicators. The details of the table provide only some of many examples of such actions. We illustrate their usefulness and importance with respect to the last column "Economic Resilience."

- Immediate (< 72 hours)—It is important to maintain a supply of critical goods and services such as water, power, and food to support the economy and social system.
- Emergency (3-7 days)—It is necessary for businesses, households, government, and nongovernment organizations to prioritize the use of resources, such as by the use of rationing. In many instances it is important to find substitutes for key inputs and to conserve them as well.
- Very Short-run (7-30 days)—The marketplace is an important inherent resource in addressing resilience. Prices reflect value and act as indicators of the scarcity of goods and services. When markets are working effectively, these price signals need to be considered in making decisions regarding the allocation of resources. When markets are not working effectively, as when market institutions are destroyed or prone to various types of market failure (including price gouging), it may be necessary for authorities to override market signals and make decisions with other

TABLE 2.1 Resilience Applications to Social, Ecological, Physical, and Economic Recovery by Time Period

Timescale	Emergency Response	Health & Safety	Utilities	Buildings	Environmental/ Ecological	Economic
Immediate < 72 hours	Tactical emergency response	Deal with casualties/ Reunite families	Use of emergency backup systems	Remove debris	Limit further ecological damage	Maintain supply of critical goods and services
Emergency 3-7 days	Strategic emergency response	Provide mass care	Begin service restoration	Provide shelter for homeless	Remove debris	Prioritize use of resources/ substitute inputs/ conserve
Very short 7-30 days	Selective response	Fight infectious outbreaks	Continue restoration	Provide shelter for homeless	Protect sensitive ecosystems	Shore up or over-ride markets
Short 1-6 months	Assist in recovery	Deal with post-traumatic stress	Complete service restoration	Provide temporary housing and business sites	Deal with ensuing problems	Cope with small business strain
Medium 6 months– 1 year	Reassess for future emergencies	Deal with post-traumatic stress	Reassess for future emergencies	Provide temporary housing and business sites	Initiate remediation	Cope with large business strain/ recapture lost production
Long >1 year	N/A	Reassess for future emergencies	Mitigation for future events	Rebuild and mitigate	Mitigate for future events	Cope with business failures/mitigation

approaches. This may be the case especially where equity, or fairness, is concerned. Free markets are known to lead to the efficient allocation of resources but are effectively blind to equity concerns.

• Short-run (1-6 months)—Small businesses are especially vulnerable in the immediate aftermath of a major disaster, and require special attention.

• Medium-run (6 months-1 year)—One of the major sources of resilience is the ability to recapture lost revenue after the event; many businesses have standing orders for their product production, and these can be filled by working overtime or extra shifts at the relatively low cost of overtime pay.

• Long-run (> 1 year)—It is important that mitigation be integrated into the reconstruction effort to reduce losses from future events.

DIMENSIONS OF RESILIENCE

Many of the points of this chapter can be reiterated by summarizing the many dimensions of resilience:

1. *Multi-scale dimension.* The concept of resilience is applicable at multiple scales, from the resilience of an individual person (e.g., psychological, financial) to that of an organization, neighborhood, city, or nation.

2. *Multi-hazard dimension.* Resilience pertains to all hazards and not just earthquakes. Moreover, resilience to other hazards can in many cases be applied to earthquakes.

3. *Stock* (property damage) *and flow* (production of goods and services) *dimensions* of assets, systems, economies, and communities. Property damage takes place at a given point in time, but the service flows (to which maintaining function applies) are disrupted until recovery is completed, and are thus more central to the idea of rebounding after a disaster.

4. *Behavioral and policy dimensions.* The length of the recovery following disasters is not some constant that can be known beforehand, but an outcome that depends critically on decisions and activities undertaken by private- and public-sector decision-makers.

5. *Geophysical dimension.* Resilience generally varies inversely to the size of the shock to the system.

6. *Bifurcation of temporal dimensions.* Static resilience refers to the ability of an entity or system to maintain function when shocked and relates to how to efficiently allocate the resources remaining after the disaster. Dynamic resilience refers to the speed at which an entity or system recovers from a shock and is a relatively more complex problem because it involves a long-term investment associated with repair and reconstruction.

7. *Contextual dimension.* The level of function of the system at a point in time has to be compared to the level that would have existed had the ability been absent, requiring that a reference point or worst-case outcome be established first.

8. *Capacity dimension.* *Inherent* resilience refers to the ordinary ability already in place to deal with crises. *Adaptive* resilience refers to ability in crisis situations to maintain function on the basis of ingenuity or extra effort.

9. *Market dimension.* This refers to the need to consider both the providers and customers of building and infrastructure services in moving toward a holistic definition of resilience.

10. *Cost dimension.* Resilience essentially represents a measure of benefits of various actions. However, the cost side cannot be neglected in policy decisions.

11. *Process dimension.* Resilience is not just about actions and targets, but the manner in which these are achieved is a critical aspect. This refers to developing and applying a set of adaptive capacities.

12. *Fairness dimension.* Resilience should be applied in an equitable manner, to be sensitive to the needs of the most disadvantaged groups in society with care being taken to try to avoid having any group adversely affected by its implementation.

3

Elements of the Roadmap

The task statement for this study charges the committee to develop a roadmap built on the goals and objectives of the 2008 NEHRP Strategic Plan. In this context, a roadmap is a plan that describes the actions and activities that will be needed to achieve the plan's objectives. Further, the charge requires an estimate of costs, recognizing that some activity costs can be specified fairly accurately (e.g., based on previous detailed studies), whereas others can only be estimated roughly. Also, some activities are scalable, that is, they can be conducted at varying levels of effort or units.

At the outset of its work, the committee was briefed on the NEHRP Strategic Plan and subsequently reviewed the plan, with supporting documents, in detail. The committee then considered the steps that would be required to make the nation and its communities more resilient to the impacts of earthquakes, based on the collective expertise of committee members and taking into account the substantial input from a community workshop (see Appendix D), but without constraining its thinking to the specifics of the Strategic Plan. In the end, 18 broad, integrated tasks, or focused activities, were identified as the elements of a roadmap to achieve earthquake resilience. These tasks are focused on specific outcomes that could be achieved in a 20-year timeframe, with substantial progress realizable within 5 years. We consider these tasks to be critical to achieving a nation of more earthquake-resilient communities.

Although the committee did not set out to explicitly ratify the elements of the Strategic Plan, in the end the committee embraced and supported these elements. The goals address loss reduction by expanding knowledge, developing enabling technologies, and applying them in vulnerable communities. The objectives identify the logical elements in fulfilling these goals.

The committee endorses the 2008 NEHRP Strategic Plan, and identifies 18 specific task elements required to implement that plan and materially improve national earthquake resilience.

The tasks identified are:

1. Physics of Earthquake Processes
2. Advanced National Seismic System
3. Earthquake Early Warning
4. National Seismic Hazard Model
5. Operational Earthquake Forecasting
6. Earthquake Scenarios
7. Earthquake Risk Assessments and Applications
8. Post-earthquake Social Science Response and Recovery Research
9. Post-earthquake Information Management
10. Socioeconomic Research on Hazard Mitigation and Recovery
11. Observatory Network on Community Resilience and Vulnerability
12. Physics-based Simulations of Earthquake Damage and Loss
13. Techniques for Evaluation and Retrofit of Existing Buildings
14. Performance-based Earthquake Engineering for Buildings
15. Guidelines for Earthquake-Resilient Lifeline Systems
16. Next Generation Sustainable Materials, Components, and Systems
17. Knowledge, Tools, and Technology Transfer to/from the Private Sector
18. Earthquake-Resilient Community and Regional Demonstration Projects

The tasks generally cross cut the goals and objectives described in the 2008 NEHRP Strategic Plan because they are formulated as coherent activities that span from knowledge building to implementation. The linkage between the goals and objectives, on the one hand, and the tasks on the other, is shown in the following matrix (Table 3.1). The matrix is richly populated, illustrating the cross-cutting nature of the tasks.

Each of the 18 tasks is described below under a series of subheadings: proposed activity and actions, existing knowledge and current capabilities, enabling requirements, and implementation issues.

TASK 1: PHYSICS OF EARTHQUAKE PROCESSES

Goal A of the 2008 NEHRP Strategic Plan is to "improve understanding of earthquake processes and impacts." Earthquake processes are difficult to observe; they involve complex, multi-scale interactions of matter and energy within active fault systems that are buried in the solid, opaque earth. These processes are also very difficult to predict. In any particular region, the seismicity can be quiescent for hundreds or even thousands of years and then suddenly erupt as energetic, chaotic cascades that rattle through the natural and built environments. In the face of this complexity, research on the basic physics of earthquake processes and impacts offers the best strategy for gaining new knowledge that can be implemented in mitigating risk and building resiliency (NRC, 2003).

The motivation for such research is clear. Earthquake processes involve the unusual physics of how matter and energy interact during the extreme conditions of rock failure. No theory adequately describes the basic features of dynamic rupture and seismic energy generation, nor is one available that fully explains the dynamical interactions within networks of faults. Large earthquakes cannot be reliably and skillfully predicted in terms of their location, time, and magnitude. Even in regions where we know a big earthquake will eventually strike, its impacts are difficult to anticipate. The hazard posed by the southernmost segment of the San Andreas Fault is recognized to be high, for example—more than 300 years have passed since its last major earthquake, which is longer than a typical interseismic interval on this particular fault. Physics-based numerical simulations show that, if the fault ruptures from the southeast to the northwest—toward Los Angeles—the ground motions in the city will be larger and of longer duration, and the damage will be much worse, than if the rupture propagates in the other direction (Figure 3.1). Earthquake scientists cannot yet predict which way the fault will eventually go, but credible theories suggest that such predictions might be possible from a better understanding of the rupture process. Clearly, basic research in earthquake physics will continue to extend the practical understanding of seismic hazards.

Proposed Actions

To move further toward NEHRP Goal A and improve the predictive capabilities of earthquake science, the National Science Foundation (NSF) and the U.S. Geological Survey (USGS) should strengthen their current research programs on the physics of earthquake processes. Bolstering research in this area will "advance the understanding of earthquake phenomena and generation processes," which is Objective 1 of the 2008 NEHRP Strategic Plan. Many of the outstanding problems can be grouped into four general research areas:

TABLE 3.1 Matrix Showing Mapping of the 18 Tasks Identified in This Report Against the 14 Objectives in the NEHRP Strategic Plan (NIST, 2008)

Task	A. Improved Understanding—Processes and Impacts					B. Develop Cost-Effective Measures to Reduce Impacts					C. Improve Community Resilience			
	1. Advance understanding of earthquake phenomena and generation processes	2. Advance understanding of earthquake effects on the built environment	3. Advance understanding of the social, behavioral, and economic factors linked to implementing risk reduction and mitigation strategies in the public and private sectors	4. Improve post-earthquake information acquisition and management	5. Assess earthquake hazards for research and practical application	6. Develop advanced loss estimation and risk assessment tools	7. Develop tools that improve the seismic performance of buildings and other structures	8. Develop tools that improve the seismic performance of critical infrastructure	9. Improve the accuracy, timeliness, and content of earthquake information products	10. Develop comprehensive earthquake risk scenarios and risk assessments	11. Support development of seismic standards and building codes and advocate their adoption and enforcement	12. Promote the implementation of earthquake-resilient measures in professional practice and in private and public policies	13. Increase public awareness of earthquake hazards and risks	14. Develop the nation's human resource base in earthquake safety fields
1. Physics of Earthquake Processes	✓	✓		✓	✓				✓	✓				
2. Advanced National Seismic System	✓	✓		✓	✓				✓					✓

Element														
3. Earthquake Early Warning	✓		✓		✓	✓					✓		✓	✓
4. National Seismic Hazard Model	✓	✓			✓	✓			✓	✓				✓
5. Operational Earthquake Forecasting	✓	✓				✓				✓	✓			✓
6. Earthquake Scenarios		✓	✓	✓	✓									✓
7. Earthquake Risk Assessment and Applications	✓	✓	✓	✓	✓	✓			✓	✓	✓	✓		✓
8. Post-Earthquake Social Science Response and Recovery Research	✓	✓			✓	✓	✓				✓	✓	✓	✓
9. Post-Earthquake Information Management	✓	✓	✓	✓	✓	✓	✓	✓	✓	✓	✓	✓		✓
10. Socioeconomic Research on Hazard Mitigation and Recovery	✓	✓	✓	✓	✓	✓			✓	✓	✓	✓	✓	✓

continued

TABLE 3.1 Continued

Task	A. Improved Understanding—Processes and Impacts				B. Develop Cost-Effective Measures to Reduce Impacts				C. Improve Community Resilience					
	1. Advance understanding of earthquake phenomena and generation processes	2. Advance understanding of earthquake effects on the built environment	3. Advance understanding of the social, behavioral, and economic factors linked to implementing risk reduction and mitigation strategies in the public and private sectors	4. Improve post-earthquake information acquisition and management	5. Assess earthquake hazards for research and practical application	6. Develop advanced loss estimation and risk assessment tools	7. Develop tools that improve the seismic performance of buildings and other structures	8. Develop tools that improve the seismic performance of critical infrastructure	9. Improve the accuracy, timeliness, and content of earthquake information products	10. Develop comprehensive earthquake risk scenarios and risk assessments	11. Support development of seismic standards and building codes and advocate their adoption and enforcement	12. Promote the implementation of earthquake-resilient measures in professional practice and in private and public policies	13. Increase public awareness of earthquake hazards and risks	14. Develop the nation's human resource base in earthquake safety fields
11. Observatory Network on Community Resilience and Vulnerability		✓	✓	✓									✓	✓
12. Physics-Based Simulations of Earthquake Damage and Loss	✓	✓			✓	✓	✓	✓		✓				✓

	1	2	3	4	5	6	7	8	9	10	11	12	13	14
13. Techniques for Evaluation and Retrofit of Existing Buildings	✓				✓			✓					✓	
14. Performance-Based Earthquake Engineering for Buildings	✓	✓		✓	✓				✓			✓	✓	✓
15. Guidelines for Earthquake-Resilient Lifeline Systems	✓		✓		✓	✓	✓		✓		✓		✓	
16. Next Generation Sustainable Materials, Components, and Systems	✓						✓	✓						
17. Knowledge, Tools, and Technology Transfer to/from the Private Sector		✓	✓							✓				
18. Earthquake-Resilient Community and Regional Demonstration Projects	✓	✓	✓	✓	✓	✓						✓	✓	

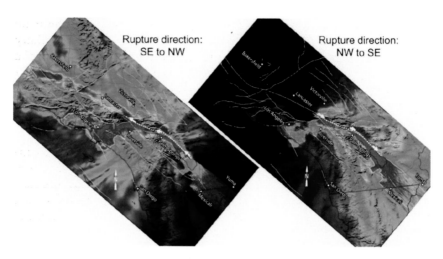

FIGURE 3.1 Maps of Southern California showing the ground motions predicted for a magnitude-7.7 earthquake on the southern San Andreas Fault; high values of shaking are purple to red, low values blue to black. The left panel shows faulting that begins at the southeast end and ruptures to the northwest. The right panel shows faulting that begins at the northwest end and ruptures to the southeast. The ground motions predicted in the Los Angeles region are much more intense and have longer duration in the former case. SOURCE: Courtesy of K. Olsen and T.H. Jordan.

• **Fault system dynamics: how tectonic forces evolve within complex fault networks over the long term to generate sequences of earthquakes.** The tectonic forces that drive earthquakes are still poorly understood. They cannot be directly measured and are influenced by unknown heterogeneities within the seismogenic upper crust as well as by slow deformation processes. The latter include intriguing new discoveries—aseismic transients such as "silent earthquakes," as well as newly discovered classes of episodic tremor and slip. How these slow phenomena are coupled to the earthquake cycle is currently unknown; a better understanding could potentially provide new types of data for improving time-dependent earthquake forecasting. A major issue is how the distribution of large earthquakes depends on the geometrical complexities of fault systems, such as fault bends, step-overs, branches, and intersections. In many cases, fault segmentation and other geometrical irregularities appear to control the lengths of fault ruptures (and thus earthquake magnitude), but large ruptures often break across segment boundaries and branch to and from subsidiary faults. For example, the magnitude-7.9 Denali earthquake in Alaska initiated as a rupture on the Susitna Glacier Thrust Fault; the rupture branched onto the main strand of the Denali Fault, and then branched again onto the subsidiary Totschunda Fault. A key objective is to develop numerical models of a brittle fault system that can simulate earthquakes over many cycles of stress accumulation and fault rupture for the purpose of constraining the earthquake probabilities used in time-dependent forecasts (see Task 5). An example of a sequence of earthquakes on the San Andreas Fault system from such an "earthquake simulator" is shown in Figure 3.2.

- **Earthquake rupture dynamics: how forces produce fault ruptures and generate seismic waves during an earthquake.** The nucleation, propagation, and arrest of fault ruptures depend on the stress response of rocks approaching and participating in failure. In these regimes, rock behavior can be highly nonlinear, strongly dependent on temperature, and sensitive to minor constituents such as water. A major problem is to understand how the microscopic processes of fault weakening control the dynamics of rupture. Are mature faults statically weak because of their compositions and elevated pore pressures, or are they statically strong but slip at low average shear stress because of dynamic weakening during rupture? Many potential weakening mechanisms have been identified—

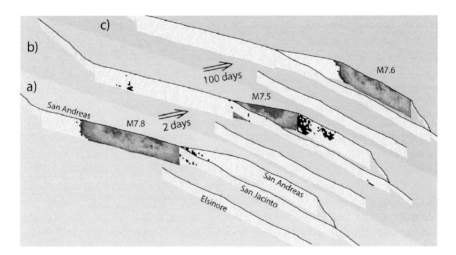

FIGURE 3.2 Example output from an earthquake simulator showing a sequence of large earthquakes during a 4-month period on the southern San Andreas Fault. There were 72 aftershocks in the 2-day interval between the magnitude-7.8 and magnitude-7.5 events, and 183 aftershocks during the 100-day interval before the magnitude-7.6 event. The three snapshots displayed here were part of a longer simulation that included 227 earthquake greater than magnitude-7. Of these, 137 were isolated by at least 4 years; 34 were pairs, and 5 were triplets such as this one. SOURCE: Courtesy of J. Dieterich.

the thermal pressurization of pore fluids, thermal decomposition, flash heating of contact asperities, partial melting, elasto-hydrodynamic lubrication, silica gel formation, and normal-stress changes due to bimaterial effects—but the physics of these processes, and their interactions, remains poorly understood. A combination of better laboratory experiments, field observation of exhumed faults, and numerical models will be required, including studies of how ruptures propagate along geometrically complex faults with distributed damage zones and off-fault plastic deformation. A priority is to validate models for application in ground motion forecasting (see Tasks 4 and 5).

•   **Ground motion dynamics: how seismic waves propagate from the rupture volume and cause shaking at sites distributed over a strongly heterogeneous crust.** Seismic hazard analysis currently relies on empirical attenuation relationships to account for event magnitude, fault geometry, path effects, and site response. These generic relationships do not adequately represent the physical processes that control ground shaking: rupture complexity and directivity, basin effects, the role of small-scale crustal heterogeneity, and the nonlinear response of the surface layers (such as soft soils). Physics-based numerical simulations of the generation and propagation of seismic radiation have now advanced to the point where they are becoming useful in predicting the strong ground motions from anticipated earthquake sources (e.g., Figure 3.1). The physics needs to account for the complexities of rupture propagation along the fault, wave propagation through the heterogeneous crust, response of the surface rocks and soils, and response of the buildings embedded in those soils. An important objective is to couple numerical models of these physical processes in end-to-end ("rupture-to-rafters") earthquake simulations (see Task 12).

•   **Earthquake predictability: the degree to which the future occurrence of earthquakes can be determined from the observable behavior of earthquake systems.** Because earthquakes cannot be deterministically predicted, forecasting requires a probabilistic (i.e., statistical) characterization of earthquake sources in terms of space, time, and magnitude (Jordan et al., 2009). Long-term earthquake forecasting is the basis for seismic hazard analysis. Current forecasts, such as those used in all three iterations of the National Seismic Hazard Maps (Frankel et al., 1996, 2002; Petersen et al., 2008), are time-independent; i.e., they assume earthquakes occur randomly in time and are independent of past seismic activity. This assumption is known to be false—almost all large earthquakes have many aftershocks, some of which can be damaging, and they often occur in clustered sequences. For example, the three largest earthquakes in the historical record of the central United States—each magnitude-7.5 or larger—occurred in the New Madrid region between mid-December, 1811, and mid-February, 1812, within a period of just 2 months.

Time-dependent forecasts that account for the occurrence of past earthquakes using stress renewal models have been developed for California (see Figure 3.10 under Task 5). However, according to these long-term models, large earthquakes on major faults decrease the probability of additional events on that fault, and they cannot therefore adequately represent the increased probability of event sequences, such as New Madrid or the hypothetical sequence illustrated in Figure 3.2. The goal of research on earthquake predictability is to develop a consistent set of probabilistic models that span the full range of forecasting timescales, long-term (centuries to decades), medium-term (years to months), and short-term (weeks to hours). Bridging the current gap between the long-term renewal models such as the Uniform California Earthquake Rupture Forecast–Version 2 UCERF2 (see Task 5) and short-term models based on triggering and clustering statistics, such as the USGS Short-Term Earthquake Probability (STEP) forecast for California[1] (Gerstenberger et al., 2007; see Figure 3.11 under Task 5), will require a better understanding of how earthquake probabilities depend on the quasi-static stress transfer caused by permanent fault slip and related relaxation of the crust and mantle, as well as the dynamic stress triggering caused by the passage of seismic waves.

Many of the potential advances in earthquake forecasting, seismic hazard characterization, and dissemination of post-earthquake information will depend on harnessing the predictive power of earthquake physics.

• Physics-based earthquake simulations can be used as tools to improve the rapid delivery of post-earthquake information for emergency management and to enable the new technology of earthquake early warning (Task 3).

• Ground motion dynamics can be used to transform long-term seismic hazard analysis into a physics-based science that can characterize earthquake hazard and risk with better accuracy and geographic resolution (Task 4).

• Research on earthquake predictability can yield better models for operational earthquake forecasting, which can help communities live with natural seismicity and prepare for potentially destructive earthquakes (Task 5).

Taken together, the technologies of Tasks 3-5 can deliver timely information needed to improve societal resilience during all phases of the earthquake cascade (Figure 3.3).

Research on earthquake physics can also contribute directly to four other NEHRP objectives. Better dynamical models of earthquake ruptures

---

[1] See earthquake.usgs.gov/earthquakes/step.

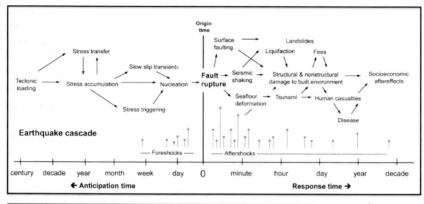

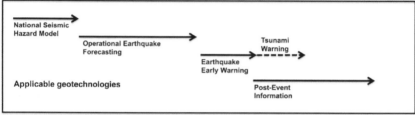

FIGURE 3.3 A diagram of the earthquake cascade showing the time domains of four geotechnologies that can improve earthquake resilience (as described under Tasks 3, 4, 5, and 8). A better understanding of earthquake physics will be needed to implement and improve these technologies. SOURCE: Courtesy Southern California Earthquake Center.

and seismic wave propagation can advance the understanding of earthquake effects on the built environment (Objective 2). A better understanding of earthquake predictability can guide the development of improved forecasting models needed for assessing earthquake hazards for research and practical application (Objective 5). Physics-based models capable of tracking earthquake cascades in real time can be used to improve the accuracy, timeliness, and content of earthquake information products (Objective 9). And more accurate earthquake simulations can provide a physical basis for developing comprehensive earthquake risk scenarios and risk assessments (Objective 10).

Existing Knowledge and Current Capabilities

Since its inception in 1977, NEHRP has been organized to gain new knowledge about earthquake hazards and risks and to implement this knowledge through effective risk mitigation and rapid earthquake response. Some of the advances in understanding earthquake processes have been highlighted in the 2008 NEHRP Strategic Plan (see Chapter 1 and Appendix A), as well as in the Earthquake Engineering Resaerch Institute (EERI) (2003b) report (see Appendix B). The National Research Council (NRC) (2003) report, *Living on an Active Earth: Perspectives on Earthquake Science*, provides one of the most expansive treatments of earthquake science and its rise under NEHRP.

The science of earthquakes, like the study of many other complex natural systems, is still in its juvenile stages of exploration and discovery. Until recently, research was focused on two primary problems: (a) earthquake complexity and how it arises from the brittle response of the lithosphere to deep-seated forces and (b) the forecasting of earthquakes and their site-specific effects. Investigations of the first problem began with attempts to place earthquake occurrence in a global framework and contributed to the discovery of plate tectonics, while work on the second addressed the needs of earthquake engineering and led to the development of seismic hazard analysis. The historical separation between these two lines of inquiry has been narrowed by progress on dynamical models of earthquake occurrence, fault ruptures, and strong ground motions. This research has transformed the field from a haphazard collection of disciplinary activities into a more coordinated "earthquake system science" that seeks to describe seismic activity not just in terms of individual events, but also as an evolutionary process involving dynamical interactions within networks of interconnected faults. Such a system-level approach recognizes that the earthquakes are emergent phenomena that depend on a wide range of interactions, from the microscopic inner scale (frictional contact asperities breaking over microseconds) to the fault-system outer scale (regional tectonic loading and relaxation over hundreds of kilometers and thousands of years).

Much has been learned from multidisciplinary investigations coordinated in the aftermath of large earthquakes, and this experience makes clear the importance of standardized instrumental data and geologic field work. Research has been accelerated through the development of new observational and computational technologies. Subsurface imaging can now be applied with sufficient resolution to delineate the deep architecture of fault systems and the three-dimensional earth structure that controls the propagation of seismic waves. In well-studied regions of the western United States, neotectonic research has improved constraints on fault geometries and long-term slip rates, and paleoseismology has furnished an extended record of past earthquakes (McCalpin, 2009), providing evidence for the clustering of large events in "seismic storms." The Global Positioning System (GPS) and Interferometric Synthetic Aperture

Radar (InSAR) satellites are mapping, with unprecedented resolution, the crustal deformations associated with individual earthquakes, long-term tectonic loading, and the stress interactions among nearby faults. Networks of broadband seismometers have been deployed to record earthquake ground motions faithfully at all frequencies and amplitudes (see Figure 3.4 under Task 2). By using high-performance computing and communications, scientists now have the means to process massive streams of observations in real time and, through numerical simulation, to quantify the many aspects of earthquake physics that have been resistant to standard analysis. New discoveries include slow slip transients that propagate at velocities systematically lower than ordinary fault ruptures.

Large earthquakes can be forecast on timescales of decades to centuries by combining the information from the geological record with data from seismic and geodetic monitoring (see Figure 3.10 under Task 5). Earthquake scientists have begun to understand how geological complexity controls the strong ground motion during large earthquakes (Figure 3.1) and, working with engineers, how to predict the site-specific response of buildings, lifelines, and critical facilities to seismic excitation. The long-term expectations for potentially destructive shaking have been quantified in the form of seismic hazard maps, which display estimates of the maximum shaking intensities expected at each locality in the United States (see Figure 3.8 under Task 4). Once a large earthquake has occurred, automated systems can rapidly and accurately compute hypocenter location, fault-plane orientation, and other source parameters. Predicted distributions of the extent of strong ground motions can be broadcast in near real time, helping to anticipate damage and guide emergency response (e.g., Figure 3.1). In the case of distant, sub-oceanic earthquakes, post-event predictions of the earthquake-generated sea waves (tsunamis) can warn coastal communities with sufficient lead times to permit evacuations. All of these advances have benefited from NEHRP-sponsored research in earthquake physics.

Enabling Requirements

Knowledge of earthquake processes is highly data limited, and there is an urgent and continuing need for better observations of earthquakes, especially through remote sensing of deformation and seismicity, and detailed field-based studies of fault rupture processes. Essential observations are provided by seismology, tectonic geodesy, and earthquake geology. The general objectives recommended in NRC (2003) have not yet been achieved:

•   An Advanced National Seismic System (ANSS) capable of recording all earthquakes down to moment magnitude-3 and up to the largest anticipated magnitude with fidelity across the entire seismic bandwidth and with sufficient density to determine the source parameters of these events. The location threshold for regional networks should reach down to magnitude-1.5 in areas of high seismic risk. Full implementation of the current ANSS plan (see Task 2) would provide this capability.

•   Geodetic instrumentation for observing crustal deformation within active fault systems with high enough spatial and temporal resolution to measure all significant motions, including aseismic events and the transients before, during, and after large earthquakes. Critical new data on earthquakes are coming from the denser networks of GPS receivers and strainmeters that have been deployed since 2005 in the Plate Boundary Observatory of NSF's EarthScope Program (EarthScope, 2007). Spatial imaging of differential motions by satellite-based InSAR has demonstrated its potential for the study of fault deformation (e.g., Helz, 2005; Pritchard, 2006). However, an InSAR satellite for collecting crustal deformation data, proposed in the original EarthScope plan (NRC, 2001), has still not been launched by the United States, and as a result researchers remain dependent on data from European and Japanese satellites (Williams et al., 2010). This reinforces the importance of the planned NASA DESDynI (Deformation, Ecosystems Structure, and Dynamics of Ice) mission, proposed to launch in 2018,[2] to provide a dedicated InSAR platform optimized for studying hazards and global environmental change.

•   Programs of geologic field study to locate active faults, quantify fault slip rates, and determine the history of fault rupture over many earthquake cycles. Light Detection and Ranging (LiDAR) techniques are now capable of high-resolution topographic imaging of fault-controlled surface morphology. For example, airborne LiDAR mapping has been used to reduce by 40% the slip along the Carrizo section of the San Andreas Fault previously ascribed to the 1857 Fort Tejon earthquake (magnitude-7.9), implying a higher medium-term probability that another large earthquake will occur on this section of the fault (Zielke et al., 2010). However, LiDAR data have been collected on a synoptic scale for only a few major faults. The methods for dating rock on neotectonic timescales of hundreds to thousands of years have been greatly improved in the past decade, but well-constrained geologic slip rates are still not available for many of the faults known to be active in the United States. Again, only a few have been studied with the paleoseismic techniques needed to resolve the slip history of fault ruptures over many earthquake cycles.

---

[2] See eospso.gsfc.nasa.gov/eos_homepage/mission_profiles/show_mission.php?id=75.

Large earthquakes are rare events, and the strong motion data from them are sparse. Numerical simulations of large earthquakes in well-studied, seismically active areas are important tools for basic earthquake science, because they provide a quantitative basis for comparing hypotheses about earthquake behavior with the limited observations. Simulations are playing an increasingly crucial role in our understanding of regional earthquake hazard and risk. This convergence of basic and applied science is comparable to the situation in climate studies, where the largest, most complex general circulation models are being used to predict the hazards and risks of anthropogenic global change. Considerable computational power will be needed to fully realize this scientific transformation and put it to practical use. Earthquakes are among the most complex terrestrial phenomena, and modeling of earthquake dynamics is one of the most difficult computational problems in science. Taken from end to end, the problem comprises the loading and eventual failure of tectonic faults, the generation and propagation of seismic waves, the response of surface sites, and—in its application to seismic risk—the damage caused by earthquakes to the built environment (see Task 12). This chain of physical processes involves a wide variety of interactions, some highly nonlinear and multi-scale. For example, long-term fault dynamics is coupled to short-term rupture dynamics through the nonlinear processes of brittle and ductile deformation, which requires earthquake simulators that can span this range of scales (see Figure 3.2).

The implementation of physics-based ground-motion prediction using numerical simulations requires estimates of the three-dimensional structure of the fault network and the material properties—seismic velocities, attenuation parameters, and density distribution—within the tectonic blocks. These structures are interrelated, because material property contrasts are often governed by fault displacements. Therefore, the development of unified structural representations requires cross-disciplinary collaboration between geologists and seismologists.

The key research issues of earthquake science are true system-level problems: they require an interdisciplinary approach to understand the nonlinear interactions among many fault-system components, which themselves are often complex subsystems. Because the behavior of each fault system is contingent on its structure, earthquake studies are necessarily conducted in a system-specific context (e.g., the Cascadia subduction zone or the San Andreas transform-fault system). Therefore, a generic understanding of earthquake processes requires a synthesis of the knowledge obtained from different regions. International collaborations can promote such a synthesis by bringing together data from many fault systems around the world.

Implementation Issues

NSF and UGSG already have well-developed research programs in earthquake physics, and strengthening those programs along the lines described here poses no major implementation issues. That said, these agencies will have to work together more closely to foster highly integrated collaborations that are (1) coordinated across scientific disciplines and research institutions, (2) enabled by high-performance computing and advanced information technology, (3) capable of assimilating new theories and data into system-level models, and (4) can partner with earthquake engineering and risk-management organizations in delivering practical knowledge to society. An additional implementation issue, of course, is the need for information from major earthquakes that can only be provided by the monitoring systems described in Task 2.

TASK 2: ADVANCED NATIONAL SEISMIC SYSTEM

Seismic monitoring is vital for meeting the nation's needs for timely and accurate information about earthquakes, tsunamis, and volcanic eruptions—information to determine their locations and magnitudes and estimate their potential effects. As well as guiding response efforts, this information also provides the basis for research on the causes and effects of earthquakes. ANSS is the USGS initiative to broadly improve the monitoring and reporting of earthquakes in the United States by integrating and modernizing the prior patchwork of state, local, and academic regional seismic networks, and coupling the seismological data with a modern earthquake information center. Begun in 2000, ANSS is modernizing and expanding capabilities nationally by establishing an integrated national system of 7,100 sensors providing data to national and regional centers. ANSS provides real-time information on the distribution and intensity of earthquake shaking to emergency responders so that they can rapidly assess the full impact of an earthquake and speed disaster relief to the most heavily affected areas. ANSS also provides engineers and designers with the information they need to improve building design standards and engineering practices to mitigate the impact of earthquakes, and provides scientists with high-quality data to understand earthquake processes and solid earth structure and dynamics. After analyzing the economic benefits of seismic monitoring, NRC (2006b; p. 8) concluded that

> Full deployment of the ANSS offers the potential to substantially reduce earthquake losses and their consequences by providing critical information for land-use planning, building design, insurance, warnings, and emergency preparedness and response. In the committee's judgment, the potential benefits far exceed the costs—annualized buildings and building-related earthquake losses alone are estimated to be about $5.6

billion, whereas the annualized cost of the improved seismic monitoring is about $96 million, less than 2 percent of the estimated losses. It is reasonable to conclude that mitigation actions—based on improved information and the consequent reduction of uncertainty—would yield benefits amounting to several times the cost of improved seismic monitoring.

Proposed Actions

The rate at which ANSS was deployed was relatively modest between 2000 and 2008, but because of substantially increased investment as part of the ARRA (American Recovery and Reinvestment Act) economic stimulus, ANSS will be about 25 percent complete by the end of 2011 (Figure 3.4). By that time, ANSS will consist of more than 1,500 modern digital seismic stations, upgraded regional seismic networks, and a National Earthquake Information Center that is operated 24×7 and delivers information for emergency response to state and local officials, operators of lifeline facilities, the Federal Emergency Management Agency (FEMA), and other critical users.

Deployment of the remaining 75 percent of ANSS is a critical requirement for national resilience, reflected by the many tasks listed in this chapter that require full ANSS deployment. One of the important components of ANSS that is still needed is an expansion of the building instrumentation component to provide crucial information on how common buildings respond to earthquake shaking.

Existing Knowledge and Current Capabilities

The ANSS plans were developed through a broad consultative process that resulted in a comprehensive description of the infrastructure elements and a detailed deployment strategy (USGS, 1999). Implementation of the plan has been approved through the USGS appropriation process, with availability of funding being the only impediment to full deployment. Because the system is already partially deployed, the technical and scientific knowledge base for ANSS is fully developed and tested.

As part of its monitoring activities, ANSS includes:

- A national "backbone" network of seismological stations.
- The National Earthquake Information Center (NEIC), the central focus for analysis and dissemination of earthquake information.
- The National Strong Motion Project, to monitor and understand the effects of earthquakes on man-made structures in densely urbanized areas to improve public earthquake safety.

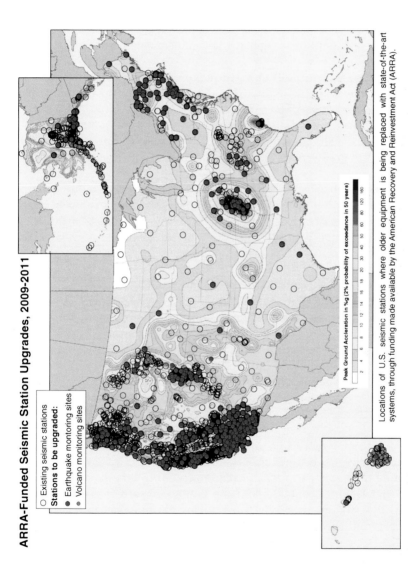

ARRA-Funded Seismic Station Upgrades, 2009-2011

Existing seismic stations
Stations to be upgraded:
Earthquake monitoring sites
Volcano monitoring sites

Peak Ground Acceleration in %g (2% probability of exceedance in 50 years)

2 4 6 8 10 12 14 16 18 20 30 40 50 60 80 120 160

FIGURE 3.4 Map showing the ANSS network of modern seismic monitoring stations either presently existing or due to be operational by the end of 2011. SOURCE: Courtesy of the USGS Earthquake Hazards Program.

Locations of U.S. seismic stations where older equipment is being replaced with state-of-the-art systems, through funding made available by the American Recovery and Reinvestment Act (ARRA).

- Fifteen regional seismic networks operated by USGS and its partners.

The range of products produced by the USGS Earthquake Hazards Program[3] derived from the ANSS network has grown steadily over recent years, as the network elements have been deployed, and these products now serve a diverse scientific, emergency management, and community base:

- **Descriptions of Recent Earthquakes.** Automatic maps and event information are available online from the Earthquake Hazards Program website within minutes of an earthquake occurring.
- *Did You Feel It* **Maps and Reports.** Present a compilation of community reports of shaking in the form of a Community Internet Intensity Map (CIIM) that summarizes the questionnaire responses provided by Internet users.
- *ShakeMaps.* Provides near-real-time maps of ground motion and shaking intensity following significant earthquakes, for use by federal, state, and local organizations, both public and private, for post-earthquake response and recovery, public and scientific information, as well as for preparedness exercises and disaster planning.
- *ShakeCasts.* Critical users, e.g., lifeline utilities, can receive automatic notifications within minutes of an earthquake, indicating the level of shaking and the likelihood of impact to their own facilities.
- **Hazard Maps.** National Seismic Hazard Maps show earthquake ground motions for various probability levels across the United States, for application in the seismic provisions of building codes, insurance rate structures, risk assessments, and other public policy uses (see Task 4).
- *PAGER* **Earthquake Notification.** Automated notifications of earthquakes through e-mail, pager, or cell phone. Rapid information and updates to first responders, and resources for media and local government.

A broad range of additional data and resources—information about earthquake hazards, historical seismicity, faults, etc. is available by state; an online searchable earthquake catalog providing downloadable information and technical data; QuickTime movies created from the recordings of fully instrumented structures during earthquakes; and real-time waveforms and spectrograms.

---

[3] See earthquake.usgs.gov/earthquakes/.

Enabling Requirements

To be fully functional, the ANSS will require the following additional components:

- **Structural instrumentation.** ANSS requirements call for extensive instrumentation of buildings, bridges, and other structures in areas of high earthquake risk. This is the least developed component of ANSS; 9,000 data channels are needed, and instrumentation installed to date is less than 1000 channels.
- **Expanded urban monitoring.** ARRA funding is targeted for the modernization of existing seismic stations, but not for an expansion of the networks. To meet the ANSS requirements, an additional 1,700 stations are needed for deployment in the highest risk urban areas.
- **Data management.** Currently, a large proportion of the data management needs of the system are being accommodated through the IRIS Data Management System, funded by NSF. At full implementation, USGS needs to assume this funding responsibility, as well as the task of developing seamless data and product access for ANSS.

Implementation Issues

Full implementation of ANSS simply requires additional funding; there are no technical issues.

TASK 3: EARTHQUAKE EARLY WARNING

ANSS, when fully implemented, will provide the infrastructure necessary for development of earthquake early warning (EEW) systems. The goal of network-based EEW is to detect earthquakes in the early stages of fault rupture, rapidly predict the intensity of the future ground motions, and warn people before they experience the intense shaking that might cause damage. The most damaging shaking is usually caused by seismic shear and surface waves, which travel at only half the speed of the fastest seismic waves, and much slower than an electronic warning message. EEW systems detect strong shaking at an earthquake's epicenter and transmit alerts ahead of the damaging earthquake waves.

Potential warning times depend primarily on the distance between the user and the earthquake epicenter. There is a "blind zone" near an earthquake epicenter where early warning is not feasible, but at more distant sites, warnings can be issued from a few seconds up to about 1 minute prior to the strong ground shaking (Figure 3.5). Such warnings can be used to reduce the harm to people and infrastructure during earth-

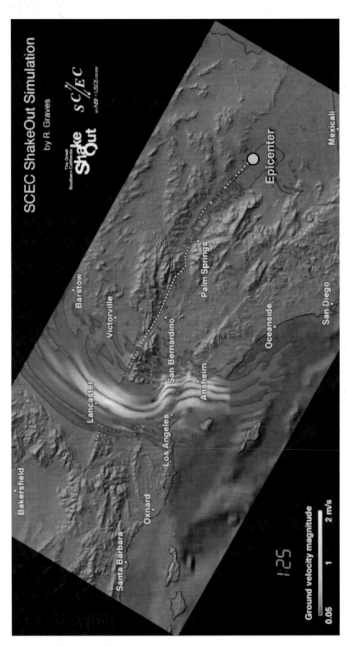

FIGURE 3.5 Snapshot of the seismic waves produced by a simulated magnitude-7.8 earthquake on the southern San Andreas Fault (dashed white line) with an epicenter at the southeastern end of the fault (yellow point). The snapshot is taken 85 seconds after the earthquake origin time, just as strong surface waves are arriving in downtown Los Angeles. In this scenario, an EEW system deployed along the southern San Andreas Fault could provide up to a minute of warning at sites in the most urbanized regions of Los Angeles. This particular earthquake simulation was used to define the hazard for the 2008 Great Southern California ShakeOut. SOURCE: Courtesy of R. Graves, G. Ely, and T.H. Jordan.

quakes. Potential applications include alerting people to "drop, cover, and hold-on," move to safer locations, or otherwise prepare for shaking (e.g., surgeons in operating rooms), as well as many types of automated actions: stopping elevators at the nearest floor, opening firehouse doors, slowing rapid-transit vehicles and high-speed trains to avoid accidents, shutting down pipelines and gas lines to minimize fire hazards, shutting down manufacturing operations to decrease potential damage to equipment, saving vital computer information to avoid losses of data, and controlling structures by active and semi-active systems to reduce building damage.

Operational EEW systems been deployed in at least five countries—Japan, Mexico, Romania, Taiwan, and Turkey. Japan is the only country with a nationwide system that provides public alerts. The Japan Meteorological Agency uses a national seismic network of about 1,000 seismological stations to detect earthquakes and issue warnings, which are transmitted via the Internet, satellite, and wireless networks to cell phone users, to desktop computers, and to automated control systems that stop trains, place sensitive equipment in a safe mode, and isolate hazards while the public takes cover (Figure 3.6). Mexico City and Istanbul also have public warning systems.

Proposed Actions

EEW has been identified as an ANSS objective (USGS, 1999), and it will be an important outcome of ANSS implementation. The NEHRP 2008 Strategic Plan recommended the evaluation and testing of EEW systems as part of Objective 9, to "Improve the accuracy, timeliness, and content of earthquake information products." Some activities are under way in California, where a USGS-sponsored demonstration project is testing several EEW algorithms using real-time data from the California Integrated Seismic Network (CISN), a component of ANSS. AARA stimulus funding is being used to upgrade many of the older seismic instruments throughout the CISN and reduce the time delays in gathering data and issuing alerts. When completed, this prototype system, called the CISN ShakeAlert System, will provide warnings to a small group of test users including emergency response groups, utilities, and transportation agencies (USGS, 2009). While in the testing phase, the system will not provide public alerts. If these tests are successful, high priority should be given to the development and deployment of an ANSS-based operational earthquake early warning system that can issue public alerts through various types of public media. The most suitable location for the first fully operational deployment of EEW would be the San Andreas Fault system, where the risk level is high and early detection of large strike-slip ruptures can provide up to a minute of early warning (e.g., Figure 3.5). If sufficient funding

Earthquake Early Warning: Dos & Don'ts

Make residences earthquake-resistant and fix furniture to prepare for earthquakes

Call the attention of those around you

If you feel a tremor   **Remain calm, and secure your personal safety!**   If you see/hear an EEW

After seeing or hearing an Earthquake Early Warning, you have only a matter of seconds before strong tremors arrive. This means you need to act quickly to protect yourself.

At Home

- Protect your head and shelter under a table
- Don't rush outside
- Don't worry about turning off the gas in the kitchen

When Driving

- Don't slow down suddenly
- Turn on your hazard lights to alert other drivers, then slow down smoothly
- If you are still moving when you feel the earthquake, pull safely over to the left and stop

In Public Buildings

- Follow the attendant's instructions
- Don't rush to the exit

Outdoors

- Look out for collapsing concrete-block walls
- Be careful of falling signs and broken glass

On Buses or Trains

- Hold on tight to a strap or a handrail

In Elevators

- Stop the elevator at the nearest floor and get off immediately

FIGURE 3.6 Portion of a leaflet prepared by the Japan Meteorological Agency describing simple instructions on how to react when an EEW alert is received. SOURCE: Japan Meteorological Agency. Available at www.jma.go.jp/jma/en/Activities/EEWLeaflet.pdf.

is made available, upgrading the prototype CISN ShakeAlert System to a fully operational, public system should be possible within 5 years.

Planning should also begin for an EEW system for the Cascadia region of the northwestern United States. Earthquakes with magnitudes greater than 8 (and as large as 9) are anticipated on the offshore megathrust of the Cascadia subduction zone. In favorable situations, EEW could provide more than a minute of warning for urban centers such as Seattle and Portland. For example, the megathrust faulting that caused the great Sumatra-Andaman Islands earthquake of December 26, 2004, (magnitude-9.2) had a total rupture duration that exceeded 1,200 seconds (Shearer and Bürgmann, 2010). Moreover, an EEW capability would complement and improve the accuracy of the tsunami warning systems already operated by National Oceanic and Atmospheric Administration (NOAA) and USGS (see NRC, 2010).

EEW systems should include the capability for enhanced alerts during periods of aftershock activity following major earthquakes, which can warn rescue personnel operating in dangerous and unstable conditions. The enhancements could be based on existing dense urban seismic networks with directed annunciation of the warning to the exposed individuals, or on fully mobile aftershock monitoring networks that can be rapidly installed in sparsely monitored locales.

Current EEW systems are based on earthquake detection and forecasting by seismometer networks such as the CISN. However, as described in the following section, continuously recording GPS networks can also provide real-time information on large fault displacements that is potentially valuable for EEW, especially in subduction zones such as Cascadia (Hammond et al., 2010). Additional research and development is needed to facilitate the rapid integration of GPS network data with seismometer network data.

Existing Knowledge and Current Capabilities

Three basic seismographic strategies have been developed for earthquake early warning (Allen et al., 2009):

- on-site or single-station warning: predicting the peak shaking from the P wave recorded at the site,
- front detection: detecting strong ground shaking at one location and transmitting a warning ahead of the seismic energy, and
- network-based warnings: using seismic networks to locate and estimate the size of a growing fault rupture.

Research indicates that dense seismometer arrays in the vicinity of shallow hypocenters can determine whether an event will grow into a large earthquake (magnitude > 6) using only several seconds of recorded P-wave data (Allen and Kanamori, 2003; Lancieril and Zollo, 2008). However, whether such measurements saturate above magnitude-7 is an unresolved problem that is related to fundamental issues of earthquake predictability.

Operational EEW systems been deployed in at least five countries—Japan, Mexico, Romania, Taiwan, and Turkey (see review by Allen et al., 2009). The most highly developed systems are in Japan. Japan Railways began using alarm-seismometers in the 1960s and then front-detection EEW systems in 1982 to shut off power to the Shinkansen bullet trains. An onsite system (Urgent Earthquake Detection and Alarm System, UrEDAS) started operation along the Shinkansen lines in 1992, which was improved after the 1995 Kobe earthquake. The system demonstrated its effectiveness during the magnitude-6.6 Niigata Ken Chuetsu earthquake of 2004, when it issued an alert that stopped a Shinkansen train. Although the train derailed, all but one car remained on the tracks. Japan has also developed a technology for network-based EEW that now provides public alerts (Kamigaichi et al., 2009). The Japan Meteorological Agency (JMA) employs a network of 1,000 seismic instruments to detect earthquakes and predict the intensity of the resulting ground motions. Warnings are sent via TV and radio and go out over public address systems in schools, some shopping malls, and train stations. Alerts of impending shaking are also transmitted via the Internet, satellite, and wireless networks to cell phone users, to desktop computers, and to automated control systems that stop trains, place sensitive equipment in a safe mode, and isolate hazards while the public takes cover (Figure 3.6). Mexico, Taiwan, Istanbul, and Bucharest have active systems providing warning to one or more users.

The finite bandwidth and the dynamic range of current seismometers limit their accuracy in measuring ground displacements near large earthquake ruptures. Complementary information can be obtained from geodetic observations using GPS networks. Continuously monitoring GPS stations can provide total displacement waveforms at sampling intervals on the order of 1 second, which can be used directly to estimate earthquake source parameters (Crowell et al., 2009). This sampling rate is lower than the seismic observations, and the noise levels of the GPS data are higher. Therefore, an integrated network of seismometers and GPS receivers can provide better performance for EEW than either type of instrument alone.

Enabling Requirements

Full implementation of ANSS, as recommended in Task 2, will provide the instrumental platform for the development of EEW systems. As noted above, development of the CISN ShakeAlert prototype is already underway. A fully operational, end-to-end system will require the densification of seismic networks in the likely epicentral regions of large earthquakes, such as along California's San Andreas Fault, which can be guided by the long-term earthquake rupture forecasts discussed under Task 4. Upgrades to the equipment currently used to record, transmit, and process seismic signals will be necessary to reduce the latencies in the automated broadcasting alerts. The robustness of the ANSS components such as the CISN will need to be improved through redundant telecommunication paths and software enhancements.

Substantial research and development will be needed on the algorithms used to detect earthquakes in the early stages of fault rupture, to predict future ground motions, and to automatically issue alerts. The basic science requirements are described under Task 1. Particularly important is a better understanding of the earthquake rupture physics, including the processes that govern the nucleation, propagation, and arrest of seismic ruptures. Short-term earthquake rupture forecasts can improve the efficacy of EEW algorithms by adjusting the a priori rupture probabilities to reflect current seismic activity (see Task 5). Automated algorithms will have to recognize and map finite-fault sources, including multi-fault ruptures, in real time.

Large uncertainties in EEW alerts of prospective shaking can arise from uncertainties in the ground motion prediction equations. The ground motion predictions will have to account for three-dimensional geologic structures, particularly near-surface heterogeneities such as sedimentary basins, and to account for rupture propagation effects such as directivity and slip complexity. Physics-based numerical simulations of strong ground motions have the potential for substantially improving these predictions (see Tasks 1 and 12).

EEW algorithms will have to be verified and validated by extensive field-testing, such as that now under way in California. This testing will need to evaluate the quality and consistency of the ground motion predictions, as well as the costs and benefits to potential users. Because of the latter requirement, the design, operation, and testing of EEW systems will have to involve end users.

Implementation Issues

Private-sector service providers will be needed to adapt EEW information for utilization in automated control and response systems. In Japan, private providers offer a variety of services ranging from simple translation of the JMA information into a site-specific predicted intensity and warning time to more sophisticated systems that incorporate local seismometers to provide additional on-site warning. Engineering and construction companies are also using the warning systems to provide both enhanced building performance during earthquakes and to protect construction workers. An effective public-private partnership will be necessary in developing "best practices" for EEW users.

Although there have been limited studies addressing the social science context of earthquake early warning (e.g., Bourque, 2001; Tierney, 2001), implementation of EEW will require additional research to determine optimal ways to interact with the public and a broad education campaign to inform the public about the availability and use of earthquake alerts. The experience of the Japanese (e.g., Figure 3.6) will be useful in this regard.

TASK 4: NATIONAL SEISMIC HAZARD MODEL

The National Seismic Hazard Maps produced by USGS are the authoritative reference for earthquake ground motion hazard in the United States. These maps are the basis of the probabilistic portion of the NEHRP Recommended Provisions, are a resource for the model building codes, and are used in seismic retrofit guidelines, earthquake insurance, land-use planning, and the design of highway bridges, dams, and landfills. They are also used in nationwide earthquake risk and loss assessment and development of credible earthquake scenarios for planning and emergency preparedness.

Proposed Actions

Improved mapping of seismic hazard, at both local and national scales, can reduce the uncertainty in earthquake probabilities and ground motion values and provide a more scientifically credible basis for engineering and policy decision-making. Seismic hazard mapping directly benefits from the advances in earthquake science described in Tasks 1, 2, and 3. Continued interaction between NEHRP researchers and the user-community will also serve to identify new earthquake hazard and risk information products of value to the community.

• **Continue the development of National Seismic Hazard Maps.** Three focus areas for the next generation of National Seismic Hazard Maps are (a) the improved characterization of faults capable of producing magnitude-6.5 to 7 earthquakes (Category B faults) using field investigations and seismic monitoring, (b) the development of improved ground motion attenuation models for the eastern and central United States, and (c) the development of, and improvements to, numerical ground motion simulations.

- **Create hazard maps for urban areas.** Expansion of the Urban Seismic Hazard Mapping program, with the goal of mapping all of the major U.S. urban areas at risk over the next 20 years. Providing greater detail about the geographic distribution of strong ground motion, geologic site conditions, and potential ground failure (fault rupture, landslides, and liquefaction) is a critical component to the earthquake risk applications discussed in Tasks 6 and 7 as well as the building and lifeline guidelines discussed in Tasks 13, 14, and 15. The development of urban seismic hazard maps involves partnerships between state and local agencies, local government, universities, and the NEHRP agencies. Integration of enhanced local hazard information with the national-scale engineering design guidance provided in the National Seismic Hazard Maps will need to be addressed by NEHRP as well as by the standards and code developing organizations. Both the San Francisco, CA, and Evansville, IN, examples described in Chapter 2 provide valuable case histories describing how such partnerships can be established.

Existing Knowledge and Current Capabilities

The current knowledge of earthquakes, active faults, crustal deformation and seismic wave generation/propagation must be integrated and translated into a form that can be used by others in order to be effective in reducing earthquake losses. The National Seismic Hazard Maps and related information products produced by USGS accomplish this critical information transfer.

Seismic Hazard Maps

During the past 60 years, the National Seismic Hazard Maps have evolved from a series of broad zones depicting 4 damage levels (none, minor, moderate, and major) based on Modified Mercalli Intensity, a qualitative measure of earthquake shaking (see Figure 3.7; Roberts and Ulrich, 1950; Algermissen, 1969), to the current series of USGS maps that provide earthquake engineering-based parameters such as spectral acceleration (Sa, at multiple periods, 0.1, 0.2, 0.3, 0.5, and 1.0 sec) and Peak Ground Acceleration (PGA) for ~150,000 sites across the country (see Figure 3.8). The current USGS hazard maps are based on a combination of state-of-the-art probabilistic methodology, ANSS earthquake monitoring, and the latest NEHRP research findings that provide a long-term geologic perspective for earthquake activity (Crone and Wheeler, 2000). These hazard maps have been developed through a scientifically defensible and repeatable process that involves input and peer review at both regional and national levels by expert and user communities (Petersen et al., 2008).

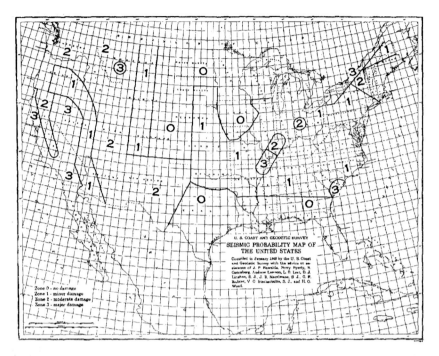

FIGURE 3.7 Seismic probability map of the United States in 1950. SOURCE: Roberts and Ulrich (1950). © Seismological Society of America.

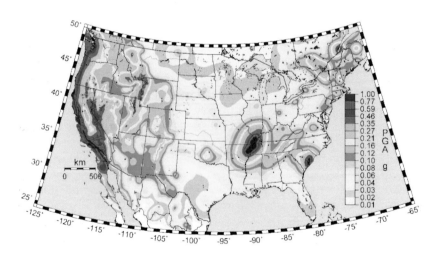

FIGURE 3.8 U.S. National Seismic Hazard Map showing Peak Ground Acceleration (PGA) with a 2 percent chance of exceedance in 50 years (or a 2,475-year return period). SOURCE: USGS (2008).

The USGS National Seismic Hazard Maps are the basis of the probabilistic portion of the NEHRP Recommended Provisions, a resource for the model building codes developed by the Building Seismic Safety Council and published by FEMA (FEMA, 2009b). These design maps are adopted by the International Building Code and national consensus standards such as ASCE-7 Minimum Design Loads for Buildings and Other Structures, ASCE-31 Seismic Evaluation of Existing Buildings, ASCE-41 Seismic Rehabilitation of Existing Buildings, and the NFPA 5000 Building Construction and Safety Code. Through these codes and standards, the National Seismic Hazard Maps affect billions of dollars of construction and represent one of the principal economic benefits of seismic monitoring in the United States (NRC, 2006b). In addition to new construction, they are used in seismic retrofit guidelines, earthquake insurance, land-use planning and the design of highway bridges (AASHTO, 2009), dams, and landfills. The national maps were used in a nationwide Hazards U.S. (HAZUS) earthquake risk assessment by FEMA (2001, 2008), and provide for basis for developing credible earthquake scenarios for planning and emergency preparedness and earthquake risk and loss assessments in the United States.

Continued NEHRP research has resulted in a new generation of earthquake hazard and risk maps that provide more specific information to support community decision-making. Urban Seismic Hazard Maps address strong ground shaking and ground failure at the community level. Seismic Risk Maps address the earthquake hazard to specific building types.

Urban Seismic Hazard Maps

Urban seismic hazard maps provide the foundation for developing realistic earthquake loss and damage estimates. By incorporating the effects of local geology, probabilistic and scenario earthquake maps provide a credible basis for community stakeholders to identify and prioritize community mitigation activities. Site and soil conditions vary geographically, and regional or local seismic hazard maps are needed to provide a higher spatial resolution to account for these differences and more accurately estimate strong ground motion effects.

A number of successful pilot programs around the United States have demonstrated the value the NEHRP Urban Seismic Hazards Mapping Program. USGS initiated a program to develop urban seismic hazard maps in 1998 for three pilot areas (San Francisco Bay region; Seattle, WA; Memphis, TN) and has since expanded the program in central United States (greater St. Louis area; Evansville, IN) and in southern California. These urban seismic hazard mapping programs involve state geological surveys, emergency management organizations, as well as local universities and consulting firms. In southern California, USGS is partnered with the Southern California Earthquake Center.

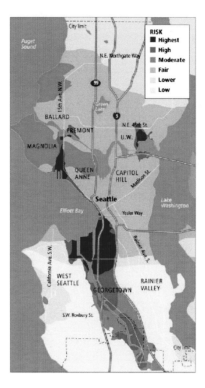

FIGURE 3.9 Seattle Urban Seismic Hazard Map, showing ground motions for the 10 percent chance of exceedance in 50 years or a 1 percent chance of exceedance in 475 years. SOURCE: Hearst Corporation. Available at seattlepi.com/U.S.G.S.

Seismic hazard maps for Seattle, WA, were improved following the magnitude-6.7 Nisqually, WA, earthquake in 2001. These maps provide a high-resolution view of potential ground shaking, which is particularly important because much of Seattle is sited on a sedimentary basin that strongly affects patterns of ground shaking and damage (see Figure 3.9). In the Nisqually earthquake, unreinforced masonry (URM) damage was disproportionately large compared with other building types, with the greatest damage occurring in areas of soft soils. Improved earthquake information for the Seattle area guided elected officials toward policy decisions about the need to mitigate hazards from URM buildings. Seattle is currently considering a URM retrofit ordinance[4] that would be the first mandatory retrofit program outside California.

---

[4] See www.cityofseattle.net/dpd/Emergency/UnreinforcedMasonryBuildings/default. asp.

Seismic Risk Maps

Earthquake risk, expressed as a level of building damage or economic loss, is dependent on both the type of building or structure and the geographic location of the structure with respect to strong ground shaking. Mapping uniform earthquake ground motions (e.g., 2 percent in 50 years, or 1 chance in 2,475 (0.04%) of exceedance in any year) does not necessarily result in identifying a uniform earthquake risk. A new series of earthquake risk maps combine hazard information from the National Seismic Hazard Maps with building fragility curves from FEMA's HAZUS-Multi-Hazard earthquake loss estimation model to show mean annual frequencies of exceeding different structural damage states (Luco and Karaca, 2007). This type of information is fundamental to seismic risk assessment (see Task 7) and can be used by communities to make risk-informed decisions and identify performance targets for specific building types based on local hazards and local building practices (e.g., 1 percent annual likelihood of collapse). Additionally, integration of this risk-map approach with USGS ShakeMaps would provide emergency responders with accurate "damage maps" for use following an earthquake impact to a risk-mapped urban area.

Enabling Requirements

The National Seismic Hazard Maps integrate knowledge of earthquakes, historic earthquakes, active faults, crustal deformation, and seismic wave generation/propagation. The availability of on-line design and analysis tools has enabled engineers and earth science professionals to determine ground motion values for specific building codes as well as create customized hazard maps. The scientific credibility of these maps is based on basic geologic and seismologic research that includes:

Earthquake Monitoring

The National Seismic Hazard maps use the basic earthquake data collected by ANSS, and, as discussed above under Task 2, ANSS is the "backbone" of seismology research in the NEHRP program.

Geologic Research

NEHRP-supported paleoseismic research has provided the necessary long-term geologic constraints on earthquake activity to validate probabilistic seismic hazard assessments. Paleoseismic information for major fault systems capable of producing earthquakes with magnitude > 7 (Category A faults such as the San Andreas, San Jacinto, Elsinore, Imperial, and

Rodgers Creek) has been well developed during the past 30 years. These techniques need to be extended to other faults lacking sufficient paleoseismic data to constrain their recurrence intervals (defined as Category B faults). Recent examples of destructive earthquakes occurring on Category B faults include the 1971 San Fernando, CA, (magnitude-6.7) and 1994 Northridge, CA (magnitude-6.7) events. In areas where time-dependent models of earthquake activity may be more appropriate, paleoseismic research on the variability of inter-event times can help identify aleatory uncertainties and help improve the overall resolution of earthquake hazard estimates.

Wave Propagation

Better ground motion attenuation models help improve structural design and construction. The introduction of the Next Generation Attenuation (NGA) models into the 2008 hazard maps (Petersen et al., 2008) modified ground motion values in many areas of the United States, significantly impacting earthquake damage and loss estimates. Continued improvement of attenuation relations for the central and eastern United States through the use of physics-based numerical simulations (see discussion in Task 1) can advance understanding of earthquake effects to the built environment and help reduce uncertainties in areas of infrequent seismicity. Significant improvements to the empirical attenuation relations may be possible through the use of numerical simulations of ground motions that incorporate realistic models of source dynamics and three-dimensional geological structure (see Figure 3.1).

Site Conditions

Active geotechnical research and mapping programs by federal, state, and local agencies, universities, and consultants continue to improve our knowledge of subsurface and geologic site effects at the community scale. The COSMOS Geotechnical Virtual Data Center[5] (Swift et al., 2004), for example, provides a distributed system for archiving and web dissemination of geotechnical data collected and stored by various agencies and organizations.

Coordination

Seismic hazard products developed by the states and university groups need to be coordinated with national maps through national and

---

[5] See www.cosmos-eq.org/.

regional peer review processes to provide nationally consistent information to users. One example is the coordination of the UCERF2 seismic hazard study maps for California with the USGS National Seismic Hazard Mapping Program (WGCEP, 2008).

Implementation Issues

As discussed in Tasks 14 and 15, predictive models of ground shaking and deformation are required for performance-based earthquake engineering. Yet, in many areas, these types of models still exhibit large uncertainties. In those regions of our nation where earthquake data are sparse or nonexistent, earthquake-physics simulations should be used to build or augment the dataset. Continued deployment of ANSS in urban environments to collect strong motion recordings and site response information is essential to validate these simulation models. Systematic expansion of hazard mapping products and the development of national- and local-scale hazard maps for liquefaction (including lateral spreading and settlement), surface fault rupture, and landslide potential is needed to complement the maps already available for ground shaking.

Although the adoption of the USGS National Seismic Hazard Maps into the model building codes is a major NEHRP success story, the actual implementation and enforcement of these codes remains a community choice. A clearer understanding by community policy-makers and stakeholders of the role that both the National Seismic Hazard Maps and the building codes play in community safety is essential for the development of earthquake-resilient communities.

TASK 5: OPERATIONAL EARTHQUAKE FORECASTING

With the current state of scientific knowledge, individual large earthquakes cannot be reliably predicted in future intervals of years or less; i.e., "deterministic" earthquake prediction is not yet possible. Nevertheless, the public needs up-to-date information about the likelihood of future events, especially following widely felt earthquakes, even if the probabilities of a strong earthquake are too small to warrant high-cost preparedness actions such as mass evacuations. The goal of operational earthquake forecasting is to provide communities with authoritative information on how seismic hazards change with time, including a consistent set of earthquake forecasts that range from the long term (centuries to decades) to the short term (hours to weeks) (Jordan et al., 2009; Jordan and Jones, 2010).

Seismic hazards are known to change on short timescales, because earthquake occurrences suddenly alter the conditions within the fault system that lead to future earthquakes. One earthquake can trigger others nearby; the probability of such triggering increases with the initial shock's magnitude and decays with elapsed time according to simple (and nearly universal) scaling laws. Statistical models of earthquake triggering can explain much of the observed spatio-temporal clustering in seismicity catalogs, such as aftershocks, and the models can be used to construct forecasts that estimate future earthquake probabilities based on prior seismic activity. These short-term models have demonstrated significant skill in forecasting future earthquakes—the probability gain factors achieved in several-day intervals can range up to 100-1,000 relative to the long-term forecasts typically used in hazard estimation described under Task 4. However, although these gain factors can be high, the forecasting probabilities for large earthquakes usually remain low in an absolute sense, rarely reaching more than a few percent for intervals of a week or less.

Nevertheless, short-term forecasts, properly applied, can be used to improve resilience. Authoritative statements about the increase in seismic hazard following a significant earthquake allow emergency management agencies, as well as the population at large, to anticipate aftershocks. Such advisories also fulfill the public's need for current information during periods of anomalous seismic activity, which can help to reduce the concern about amateur predictions and rumors that overly inflate the hazard.

Under the Stafford Act (P.L. 93-288), USGS has the federal responsibility for earthquake monitoring and forecasting. Its National Earthquake Prediction Evaluation Council (NEPEC) provides advice and recommendations on earthquake forecasts and related scientific research to the USGS director, in support of the director's delegated responsibility to issue timely warnings of potential geologic disasters. Thus far, USGS and NEPEC have not established protocols for operational forecasting on a national level.

Proposed Actions

USGS should develop a national plan, coordinated with state and local agencies, for the implementation of operational earthquake forecasting. In formulating the plan, USGS should consider the following elements:

- **Support for research.** Through its internal research program and external grants program, USGS should continue to support research on the scientific understanding of earthquakes and earthquake predictability.
- **Coordination of earthquake information.** USGS should continue to coordinate across federal and state agencies to improve the flow of earthquake information, particularly the real-time processing of seismic and geodetic data and the timely production of high-quality earthquake catalogs. Full support of ANSS operations will allow substantial improvements in the real-time seismic information needed for short-term forecasting.

- **Development of operational systems.** USGS should support the development of earthquake forecasting methods—based on seismicity changes detected by ANSS—to quantify short-term probability variations, and it should deploy the infrastructure and expertise needed to utilize this probabilistic information for operational purposes. Working with local agencies, USGS should provide the public with authoritative, scientific information about the short-term probabilities of future earthquakes. The source of this information needs to properly convey the epistemic uncertainties in these forecasts.

- **Operational qualification of forecasts.** All operational procedures involved with the creation, delivery, and utility of forecasts should be rigorously reviewed by experts. Earthquake forecasting procedures should be qualified for usage according to the three criteria commonly applied in weather forecasting (Jordan and Jones, 2010): they should display quality, a good correspondence between the forecasts and actual earthquake behavior; consistency, compatibility among procedures used at different spatial or temporal scales; and value, realizable benefits (relative to costs incurred) by individuals or organizations who use the forecasts to guide their choices among alternative courses of action.

  o Operational forecasts should incorporate the results of validated short-term seismicity models that are consistent with the authoritative long-term forecasts and demonstrate reliability (correspondence to observations collected over many trials) and skill (performance relative to the long-term forecast).

  o Verification of reliability and skill requires objective evaluation of how well the forecasting model corresponds to data collected after the forecast has been made (prospective testing), as well as checks against data previously recorded (retrospective testing). All operational models should be subject to continuous prospective testing against established long-term forecasts and a wide variety of alternative, time-dependent models.

  o Experience has shown that such evaluations are most diagnostic when the testing procedures conform to rigorous standards, and the prospective testing is blind (Field et al., 2007). In this regard, advantage can be taken of the Collaboratory for the Study of Earthquake Predictability (CSEP),[6] which has begun to establish standards and an international infrastructure for the comparative, prospective testing of earthquake forecasting models (Zechar et al., 2010). Regional experiments are now under way in California, New Zealand, Japan, and Italy, and will soon be started in China; a program for global testing has also been initiated.

---

[6] See www.cseptesting.org/.

o   Continuous testing in a variety of tectonic environments will be critical for demonstrating the reliability and skill of the operational forecasts, and for quantifying their uncertainties. At present, seismicity-based forecasts can display order-of-magnitude differences in probability gain, depending on the methodology, and there remain substantial issues about how to assimilate the data from ongoing seismic sequences into the models.

• **Assessment of forecast utility.** Most previous work on the public utility of earthquake forecasts has anticipated that they would deliver high probabilities of large earthquakes, i.e., deterministic predictions would be possible. This expectation has not been realized. Current forecasting policies need to be adapted to applications in a "low-probability environment"—one in which earthquake forecasting probabilities can vary by several orders of magnitude, but remain low in an absolute sense (< 10 percent in the short term).

The implementation of operational earthquake forecasting will enable cost-effective measures to reduce earthquake impacts on individuals, the built environment, and society-at-large—Goal B in NIST (2008). A national plan for operational forecasting will address NEHRP Objective 5 (assess earthquake hazards for research and practical application), and it will provide new information tools for Goal C (improve the earthquake resilience of communities nationwide), particularly for Objective 9 (improve the accuracy, timeliness, and content of earthquake information products).

Existing Knowledge and Current Capabilities

An up-to-date overview of existing knowledge and capabilities in earthquake forecasting and prediction is the subject of an extensive recent review by the International Commission on Earthquake Forecasting (ICEF), which was convened by the Italian government following the magnitude-6.3 L'Aquila earthquake of April 6, 2009 (Jordan et al., 2009). The statements in this section are based on this overview.

Given the current state of scientific knowledge, individual large earthquakes cannot be reliably predicted in future intervals of years or less; i.e., reliable and skillful deterministic earthquake prediction is not yet possible. In particular, the search for diagnostic precursors—signals observed before earthquakes that reliably indicate the location, time, and magnitude of an impending event—has not yet produced a successful short-term prediction scheme.

Any information about the future occurrence of earthquakes contains large uncertainties and therefore needs to be expressed in terms of probabilities. Probabilistic earthquake forecasting is a rapidly evolving field of earthquake science. Long-term forecasts provide probabilistic estimates of where earthquakes will occur, how large they might be, and how often they will happen, averaged over time intervals of decades to centuries. This information is essential for seismic hazard mapping, and it is the foundation on which the operational earthquake forecasting is built (see Task 4).

Earthquakes tend to cluster in space and time. Large earthquakes produce aftershocks by stress triggering, and sequences of earthquakes clustered in space and time are common. Aftershock excitation and decay, as well as other aspects of earthquake clustering, show statistical regularities on timescales of hours to weeks that can be captured in short-term earthquake forecasts. Additional information on earthquake probabilities over the medium-term (months to years) can be obtained from the disturbance of the tectonic forces acting on faults caused by previous large earthquakes.

Although this type of seismicity-based forecasting can provide substantial probability gains relative to long-term forecasts, the absolute probabilities remain low. Consider the southernmost segment of California's San Andreas Fault, which has a fairly high long-term probability; according to the UCERF2 model (Figure 3.10), there is a 1-in-4 chance of a magnitude ≥ 7 earthquake occurring on this fault during the next 30 years. Over a 3-day period, however, the probability of such an event is very small, about 10^{-4}. In March 2009, a swarm of more than 50 small earthquakes occurred within a few kilometers of the southern end of this fault, near Bombay Beach, California, including an magnitude-4.8 event on March 24. Using a methodology developed for assessing foreshocks on the San Andreas Fault, the California Earthquake Prediction Evaluation Council (the state equivalent of NEPEC) estimated that the swarm increased the 3-day probability of a major earthquake on the San Andreas to about 1-5 percent, corresponding to a gain factor of about 100-500 relative to UCERF2.

Foreshocks cannot be discriminated a priori from background seismicity. Worldwide, less than 10 percent of earthquakes are followed by a larger earthquake within 10 kilometers and 3 days; less than half of the large earthquakes have such foreshocks. Many earthquakes strike without warning; for example, no foreshocks or other short-term precursors have been reported for the magnitude-7 Haiti earthquake of January 12, 2010, the fifth-deadliest seismic disaster in recorded history.

Protocols for issuing advisories are best developed in California, where the dissemination of forecasting products is becoming more automated (Jordan and Jones, 2010). For every earthquake recorded above magnitude-5, the California Integrated Seismic Network, a component of the ANSS, now automatically posts the probability of an magnitude

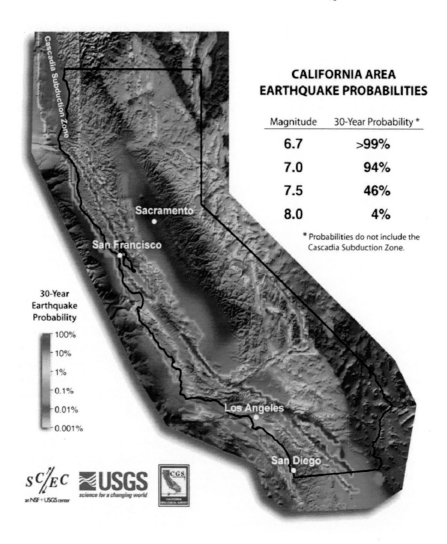

FIGURE 3.10 Uniform California Earthquake Rupture Forecast. SOURCE: Field et al. (2007); U.S. Geological Survey.

≥ 5 aftershock and the number of magnitude ≥ 3 aftershocks expected in the next week. Authoritative short-term forecasts are also becoming more widely used in other regions. For instance, beginning on the morning after the damaging L'Aquila earthquake of April 6, 2009, the Italian authorities began to post 24-hour forecasts of aftershock activity.

An operational system is the Short-Term Earthquake Probability (STEP)

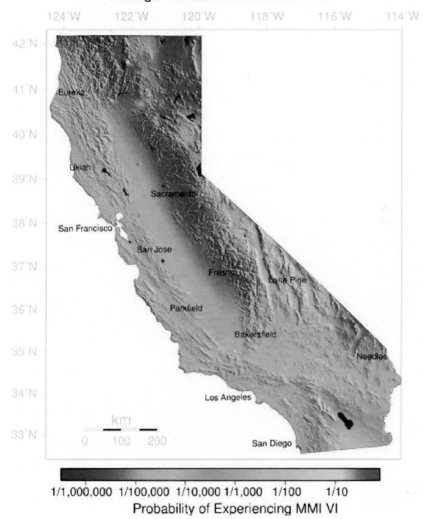

Forecast for 12/01/2010 10:00 AM PST
through 12/2/2010 10:00 AM PST

Probability of Experiencing MMI VI

FIGURE 3.11 Short Term Earthquake Probability (STEP) map. SOURCE: U.S. Geological Survey; continuously available on-line at earthquake.usgs.gov/earthquakes/step/.

model, an aftershock forecasting web service provided for California by USGS since 2005 (Gerstenberger et al., 2007). STEP uses aftershock statistics to make hourly revisions of the probabilities of strong ground motions (Modified Mercalli Intensity ≥ VI) on a 10-km, statewide grid (Figure 3.11).

Data other than seismicity have been considered in earthquake forecasting (e.g., geodetic measurements and geoelectrical signals), but so far, studies of non-seismic precursors have not quantified short-term probability gain, and they therefore cannot be incorporated into operational forecasting methodologies (Jordan et al., 2009).

Enabling Requirements

A fundamental uncertainty in earthquake forecasting is the short sampling interval available from instrumental seismicity catalogs and historical records, which is reflected in the large epistemic uncertainty in earthquake recurrence statistics. These uncertainties can be reduced by better instrumental catalogs, improved geodetic monitoring, and geologic field work to identify active faults, their slip rates, and recurrence times. ANSS implementation is an enabling requirement.

Increasing the (low) probability gains afforded by existing forecasting models will require a much improved understanding of earthquake predictability. This is an important goal of the NEHRP basic science program described under Task 1. A particular knowledge gap is our lack of knowledge about the state of stress in active fault systems and how this stress evolves over time.

Current models used for aftershock forecasting can be improved by incorporating more information about main shock deformation patterns and geological settings, such as more detailed descriptions of local fault systems. In the STEP prototype system, for example, the probability change calculated to result from a particular earthquake does not depend on the proximity of that earthquake to major faults. In this regard, short-term forecasting models that incorporate earthquake clustering and triggering need to be integrated with long-term, fault-based models, such as UCERF. A new Working Group on California Earthquake Probabilities plans to incorporate short-term forecasting into the next version of the fault-based UCERF3, which is due to be submitted to the California Earthquake Authority in mid-2012.

Forecasting models considered for operational purposes should demonstrate reliability and skill with respect to established reference forecasts, such as long-term, time-independent models. Verification of reliability and skill requires objective evaluation of how well the forecasting model corresponds to data collected after the forecast has been made (prospective testing), as well as checks against data previously recorded (retrospective testing). CSEP is setting up an infrastructure for this purpose (Zechar et al., 2009). The adaptation of CSEP to the testing of operational forecasts faces a number of conceptual and organizational issues. For example, fault-based models will need to be reformulated to permit rigorous testing—a considerable challenge for the development of UCERF3 and more advanced versions of this time-dependent California model.

CSEP evaluations are currently based on comparisons of earthquake forecasts with seismicity data. However, from an operational perspective, forecasting value can be better represented in terms of the strong ground motions that constitute the primary seismic hazard. This approach has been applied in the STEP model, which forecasts ground motion exceedance probabilities at a fixed shaking intensity, and it should be considered in the future formulation and testing of operational models. The coupling of physics-based ground motion models with earthquake forecasting models offers new possibilities for developing ground motion forecasts.

Implementation Issues

The utilization of earthquake forecasts for risk mitigation and earthquake preparedness requires two basic components—scientific advisories expressed in terms of probabilities of threatening events, and protocols that establish how probabilities can be translated into mitigation actions and preparedness. Although some experience has been gained in California (Jones et al., 1991; Jordan and Jones, 2010), there is no formal national approach for converting earthquake probabilities into mitigation and preparedness actions. One strategy that can assist decision-making is the setting of earthquake probability thresholds for such actions. These thresholds should be supported by objective analysis, for instance by cost/benefit analysis, in order to justify actions taken in a decision-making process.

Providing probabilistic forecasts to the public in a coordinated way is an important operational capability. Good information keeps the population aware of the current state of hazard, decreases the impact of ungrounded information, and contributes to reducing risk and improving preparedness. The principles of effective public communication have been established by social science research and should be applied in communicating seismic hazard information.

TASK 6: EARTHQUAKE SCENARIOS

Earthquake risk studies can take the form of deterministic or scenario studies where the effects of a single earthquake are modeled, or probabilistic studies that weigh the effects from a number of different earthquake scenarios by their annual likelihood or frequency of occurrence. Task 6 addresses the role of the individual scenario in community planning, and Task 7 addresses the earthquake risk assessment and loss estimation methodologies themselves. Earthquake scenarios integrate earth science, engineering, and social science information into a format that enables communities to visualize the impacts from earthquakes without actually having the event occur. Using scenarios, communities can evaluate potential local and regional disruptions to the built environment and society, as well as their capabilities to respond to, and recover from, earthquakes, and they can start to identify the necessary steps to reduce such impacts in the future.

Proposed Actions

The development of realistic earthquake scenario maps involves the linking of scientifically credible earthquake and ground motion maps with high-resolution urban geology and cultural inventory information in a GIS platform. Many of the issues associated with establishing credible earthquakes, ground motion, and local site condition maps are discussed under Task 5. Guidelines for the development and conduct of earthquake scenarios using NEHRP products were proposed in EERI (2006) to provide communities with information about the level of detail and effort required:

• **Development of additional scenario ShakeMaps for high-risk communities.** Currently, only a few of the 18 states with high or very high seismicity have ShakeMap scenarios readily available on the web for use in scenario and exercise development.[7] Producing a comprehensive series of ShakeMaps for all of the 43 high-risk communities identified in either USGS Circular 1188 (USGS, 1999) or FEMA 366 (FEMA, 2008) (see Table 3.2) should be undertaken by the NEHRP program during the next 5 years. ShakeMap guidelines for earthquake scenarios (Wald et al., 2001) provide technical information to assist with scenario development. Over the next 20 years, NEHRP should continue to update this information by incorporating the latest developments in both the National and Urban Seismic Hazard and Risk Maps.
• **Local data collection.** There is a widely recognized need to increase the level of detail of building and inventory data at the local level. Locally coordinated data collection can increase the resolution and

---

[7]See www.earthquake.usgs.gov/earthquakes/shakemap/list.php?y=2011 (accessed November 30, 2010).

TABLE 3.2 HAZUS-MH Annualized Earthquake Loss (AEL) and Annualized Earthquake Loss Ratios (AELR) for 43 High-Risk (AEL greater than $10 million) Metropolitan Areas

Rank	State	AEL ($ Million)	Rank	State	AELR ($/Million $)
1	Los Angeles-Long Beach-Santa Ana, CA	1,312.3	1	San Francisco-Oakland-Fremont, CA	2,049.44
2	San Francisco-Oakland-Fremont, CA	781.0	2	Riverside-San Bernardino-Ontario, CA	2,021.57
3	Riverside-San Bernardino-Ontario, CA	396.5	3	El Centro, CA	1,973.77
4	San Jose-Sunnyvale-Santa Clara, CA	276.7	4	Oxnard-Thousand Oaks-Ventura, CA	1,963.00
5	Seattle-Tacoma-Bellevue, WA	243.9	5	San Jose-Sunnyvale-Santa Clara, CA	1,837.58
6	San Diego-Carlsbad-San Marcos, CA	155.2	6	Santa Rosa-Petaluma, CA	1,662.57
7	Portland-Vancouver-Beaverton, OR-WA	137.1	7	Santa Cruz-Watsonville, CA	1,580.97
8	Oxnard-Thousand Oaks-Ventura, CA	111.0	8	Los Angeles-Long Beach-Santa Ana, CA	1,574.85
9	Santa Rosa-Petaluma, CA	68.6	9	Napa, CA	1,398.18
10	St. Louis, MO-IL	58.5	10	Vallejo-Fairfield, CA	1,375.94
11	Salt Lake City, UT	52.3	11	Anchorage, AK	1,238.56
12	Sacramento-Arden-Arcade--Roseville, CA	52.0	12	Santa Barbara-Santa Maria-Goleta, CA	1,207.93
13	Vallejo-Fairfield, CA	39.8	13	Reno-Sparks, NV	1,150.40
14	Memphis, TN-MS-AR	38.2	14	Bremerton-Silverdale, WA	1,110.13
15	Santa Cruz-Watsonville, CA	36.2	15	Salinas, CA	1,075.54
16	Anchorage, AK	34.8	16	Seattle-Tacoma-Bellevue, WA	1,052.43
17	Santa Barbara-Santa Maria-Goleta, CA	34.4	17	Salt Lake City, UT	984.61
18	Las Vegas-Paradise, NV	33.1	18	Olympia, WA	969.50
19	Honolulu, HI	32.0	19	Portland-Vancouver-Beaverton, OR-WA	942.62
20	Bakersfield, CA	30.3	20	Bakersfield, CA	870.43
21	New York-Northern New Jersey-Long Island, NY-NJ-PA	29.9	21	San Luis Obispo-Paso Robles, CA	848.65
22	Salinas, CA	29.2	22	Ogden-Clearfield, UT	826.52
23	Reno-Sparks, NV	29.0	23	Salem, OR	797.50
24	Charleston-North Charleston, SC	22.3	24	San Diego-Carlsbad-San Marcos, CA	770.20
25	Columbia, SC	21.6	25	Charleston-North Charleston, SC	766.01
26	Stockton, CA	20.9	26	Eugene-Springfield, OR	701.95
27	Atlanta-Sandy Springs-Marietta, GA	19.1	27	Provo-Orem, UT	683.30
28	Bremerton-Silverdale, WA	17.7	28	Stockton, CA	597.79
29	Ogden-Clearfield, UT	17.5	29	Memphis, TN-MS-AR	509.13
30	Salem, OR	17.4	30	Evansville, IN-KY	485.60
31	Eugene-Springfield, OR	16.5	31	Columbia, SC	478.05
32	Napa, CA	15.9	32	Modesto, CA	473.60
33	San Luis Obispo-Paso Robles, CA	15.7	33	Las Vegas-Paradise, NV	390.28
34	Nashville-Davidson--Murfreesboro, TN	15.4	34	Sacramento--Arden-Arcade--Roseville, CA	374.73
35	Albuquerque, NM	14.7	35	St. Louis, MO-IL	337.23
36	Olympia, WA	13.7	36	Albuquerque, NM	322.20
37	Modesto, CA	13.0	37	Honolulu, HI	311.12
38	Fresno, CA	12.6	38	Fresno, CA	283.13
39	Evansville, IN-KY	11.7	39	Little Rock-North Little Rock, AR	248.74
40	Birmingham-Hoover, AL	11.3	40	Nashville-Davidson-Murfreesboro, TN	167.26
41	El Centro, CA	10.7	41	Birmingham-Hoover, AL	115.54
42	Little Rock-North Little Rock, AR	10.5	42	Atlanta-Sandy Springs-Marietta, GA	65.39
43	Provo-Orem, UT	10.4	43	New York-Northern New Jersey-Long Island, NY-NJ-PA	20.90

SOURCE: FEMA (2008).

reduce the uncertainty in earthquake scenario results. This includes using local assessor databases or specialized inventories (ImageCat, Inc. and ABS Consulting, 2006) and updating those inventories using tools such as the HAZUS Comprehensive Data Management System (CDMS)[8] and the Rapid Observation of Vulnerability and Estimation of Risk (ROVER) (Porter et al., 2010) to produce upgrades to data necessary for conducting more site specific analyses.

- **Community earthquake exercises.** Community earthquake exercises provide the opportunity for communities to assemble the hazard studies and collect inventories, and stimulate community involvement, to better understand and prepare for an eventual earthquake. The success of the 2008 Great ShakeOut earthquake exercise in southern California has lead to the establishment of yearly statewide ShakeOut drills throughout California.[9] This success has led to other states adopting the ShakeOut model, including Nevada in 2010[10] and those in the central United States for the New Madrid earthquake bicentennial in 2011.[11]

Existing Knowledge and Current Capabilities

Earthquake scenarios provide opportunities to examine alternative outcomes and stimulate creative thinking about the need for new policies and programs. Incorporating the latest scientific, engineering, and societal knowledge about a region's seismic hazard, local soil characteristics, building types, lifelines, and population characteristics, a scenario can create a compelling picture that members of the local community can recognize and relate to. Scenarios show communities the potential levels of disruption of their daily life and how long the disruption may last, providing a motivation to perform the necessary actions to reduce impacts. Not only can such scenarios stimulate new policies and programs, but also the process of scenario development itself often results in greater understanding and improved trust and communication between members of the scientific, engineering, emergency management, and policy communities, resulting in a "new community" dedicated to seismic risk reduction.[12]

Earthquake scenarios have been developed for a number of fault zones in the United States, and are available from the EERI Developing Earthquake Scenarios website.[13] The earthquake scenarios that have been developed for California include the Hayward and San Andreas Faults in

---

[8] See www.fema.gov/plan/prevent/hazus/hz_cdms2.shtm.

[9] See www.shakeout.org/.

[10] See www.shakeout.org/nevada (accessed November 30, 2010).

[11] See newmadrid2011.org/.

[12] See www.eeri.org/site/projects/eq-scenarios (accessed May 4, 2010).

[13] See www.nehrpscenario.org (accessed Feb 5, 2011).

the San Francisco Bay region (CGS, 1982, 1987; EERI, 1996, 2005; Kircher et al., 2006) and the San Andreas, San Jacinto, and Newport Englewood Faults in southern California (CGS, 1982, 1988, 1993; Jones et al., 2008; Perry et al., 2008). Both the Bay Area and southern California scenarios impact some of the largest population centers in the United States, with damage estimates ranging between $100 and $200 billion and with thousands of fatalities and tens of thousands of injuries. Similarly, scenario indications that earthquake-induced levee failures in the Sacramento-San Joaquin River delta would disrupt drinking water supplies to more than 22 million Californians as well as irrigation water to delta and state agricultural lands[14] provides a powerful motivation for community awareness programs and mitigation activities.

In the Pacific Northwest, scenarios for a great Cascadia earthquake (CGS, 1995; CREW, 2005), as well as a magnitude-6.7 earthquake on the Seattle Fault (EERI, 2005), have been developed based on NEHRP research. Both the Cascadia Region Earthquake Workgroup (CREW) and the EERI reports were developed through collaborative public-private efforts involving local public- and private-sector organizations including the American Society of Civil Engineers (ASCE), Structural Engineers Association of Washington (SEAW), USGS, University of Washington, and Washington State Emergency Management (Ballantyne, 2007). The Cascadia scenario drew examples from recent great subduction zone earthquakes, such as the 1964 Alaska and 2004 Sumatra events, to illustrate some of the effects that these events would have on local communities. In contrast to the scenarios for large urban areas in California, the magnitude-6.7 Seattle Fault scenario provided a small city perspective (Figure 3.12). Damage to modern and older construction was estimated at $33 billion, with 1,600 fatalities. Focus on heavily impacted areas, such as Pioneer Square—which was badly damaged in the 2001 Nisqually earthquake—provided additional realism and credibility to the scenario.

The Oregon Department of Geology and Mineral Industries (DOGAMI) worked with Oregon Emergency Management and the University of Oregon to develop countywide earthquake and landslide hazard maps as well as earthquake damage and loss estimates as part of its natural hazard mitigation plans. Based on improved information, one Cascadia earthquake scenario estimates more than $11 billion in building damages for the mid- and southern Willamette Valley (Burns et al., 2008).

In the central United States, the FEMA New Madrid Catastrophic Planning Initiative is being developed for the 200th anniversary of the 1811/1812 New Madrid earthquakes, involving 4 FEMA regions, 8 states, and detailed assessments in 161 counties in the 8 states (Alabama, Arkan-

---

[14] See www.water.ca.gov/news/newsreleases/2005/110105deltaearthquake.pdf.

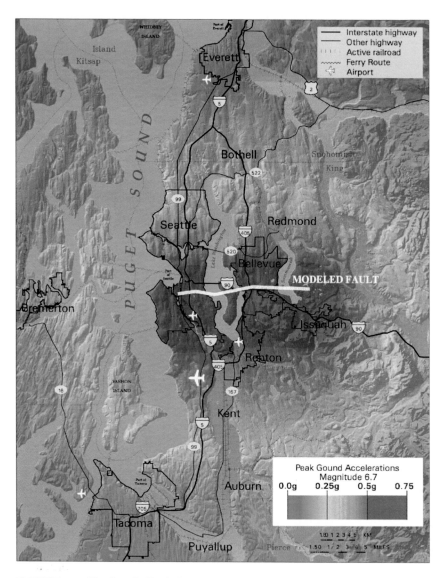

FIGURE 3.12 The Seattle Fault Scenario, depicting the impact of a magnitude-6.7 earthquake on the Seattle Fault. SOURCE: Weaver et al. (2005); adapted from U.S. Geological Survey.

sas, Illinois, Indiana, Kentucky, Mississippi, Missouri, and Tennessee). The project created new, regionally comprehensive soil characterization maps, new ground motion maps for scenario events, updated transportation and utility networks models for Memphis, TN, and St. Louis, MO, and methods to quantify the uncertainty in various impact model results. Initial scenario results for a 2:00 a.m. scenario include 3,500 fatalities, 86,000 injured, ~$300 billion in direct economic loss, ~715,000 damaged houses, and ~2.6 million households without electrical power (Elnashai et al., 2009; see Box 2.1). In the eastern United States, an earthquake loss estimation for the metropolitan New York–New Jersey–Connecticut area showed that even a moderate earthquake would significantly impact the region's large population (18.5 million) and predominately unreinforced masonry building stock (Tantala et al., 2003). South Carolina recently completed a comprehensive risk assessment for the repeat of the 1886 magnitude-7.3 Charleston earthquake, producing an estimate of $20 billion in direct losses (URS et al., 2001).

Earthquake scenarios also provide situational awareness for emergency managers. Hawaii has developed a web-based catalog, the Hawaii HAZUS Atlas,[15] of 20 "plausible" hypothetical earthquakes based on historic events that have happened in and around Maui and Hawaii counties. During an actual earthquake, emergency managers would be able to quickly assess the situation using a scenario event with similar location and size, while HAZUS modelers at the Pacific Disaster Center analyze the actual earthquake in near-real time and issue event-specific information.

Enabling Requirements

Scientifically credible earthquake scenarios and ground motions are based on NEHRP products such as the National Seismic Hazard Maps and ShakeMaps. Disaggregation of the national hazard maps to produce scenario ground motion maps allows communities to examine the local seismic hazard from individual earthquakes. The maps are usually produced for peak ground acceleration, peak velocity, and acceleration at various periods, which would affect structures of different heights or lengths. Urban Seismic Hazard Maps integrate the necessary information about geologic hazards and characteristics at the community level. High-resolution local information is used to refine bedrock ground motion inputs and ground failure models. Levels of shaking at the ground surface depend on the thickness and nature of the soils resting on the bedrock. These types of data are captured through urban hazard mapping projects such as the pilot programs in the eastern San Francisco Bay area; Seattle,

---

[15] See www.pdc.org/hha.

WA; Memphis, TN; Evansville, IN; and the Greater St. Louis area. Urban Risk Maps, coupled with HAZUS-MH loss estimation software and databases, enable economic and social loss estimates as well as physical damage estimates. All these parameters combine to provide a realistic picture of the various types of impacts earthquake can have on a local community.

Implementation Issues

Federal, state, and local emergency management organizations provide the framework to conduct and organize community exercises. Although pilot studies have demonstrated the value of earthquake scenarios for increasing public awareness, these are ultimately community-level programs where success is dependent on the extent of community involvement. Communities need to feel that they "own" the scenarios and the results that stem from these exercises, and increasing local capacity by providing support and training for staff helps to establish that ownership. FEMA-sponsored HAZUS training, coupled with guideline development and networking support for scenario developers through professional organizations like EERI, provides communities with the tools and the capabilities to develop their own scenarios. Scenarios can also allow stakeholders to perform "what if" types of analyses (i.e., if we mitigate x what is the benefit to y?) to help identify cost-effective mitigation and loss avoidance strategies.

TASK 7: EARTHQUAKE RISK ASSESSMENTS AND APPLICATIONS

While national seismic hazard maps and earthquake scenarios contribute to understanding earthquake hazards, there is an increased recognition among policy-makers, researchers, and practitioners of the need to analyze and map earthquake risk in the United States. As urban development continues in earthquake-prone regions, there is a growing concern about the exposure of buildings, lifelines, and people to the potential effects of destructive earthquakes. Earthquake risk assessments and loss estimations build on the scenario earthquakes described in Task 6 by integrating engineering and social science information in a GIS-based loss estimation methodology. Although publicly available risk assessment methodologies, data, and results have been developed and used by states and local communities, much has been based on simplified analysis modules and the use of estimated parameters or data. This has reduced the granularity of the analyses, creating uncertainty and limiting the ability to identify and act on specific hazard and risk issues. Many of these uncertainties can be addressed and reduced through NEHRP activities.

Proposed Actions

The primary source of uncertainty in loss estimation models is the lack of accurate input data. This includes not only the data used by the models—such as information about seismic source characterization, strong ground motion attenuation, local soil conditions, and inventories of the built environment—but also data used to develop the models themselves. Different parameters used in different loss estimation models can change the level of uncertainty of the mean by a factor of five or more. A sensitivity study conducted by Porter et al. (2002) showed that the parameters that most impact the damage estimate for a particular building are related to earthquake ground motion and include the fragility curves, which provide an estimate of damage to various building components as a function of ground motion, the spectral acceleration of the ground motion used in the analysis, and the ground motion record or time history used in the analysis. For example, one AEL estimate for California based on improved soils classification is ~30 percent less than an estimate based on a single default soil type throughout the state (Rowshandel et al., 2003; FEMA, 2008). Similarly, California building related losses based on Next Generation Attenuation (NGA) models are 28 percent to 63 percent lower than those based on earlier ground motion prediction equations (Chen et al., 2009). Modifications to the HAZUS building damage modules in the HAZUS-MH4 release have been shown in earthquake simulations in Washington State to reduce estimates of death, injuries, and the number of people requiring shelter by as much as 30 percent (Terra et al., 2010). These types of reductions reflect improvements in the ability to characterize both hazard and risk. Continued improvements to earthquake risk and loss estimation methodologies and the development of community risk models are two activities that have been identified as NEHRP focus areas:

1. Promote the continued development and enhancement of earthquake risk assessment and loss estimation methodologies and databases. EERI (2003b) identified five areas of emphasis for system level simulation and loss assessment:

- Validation studies to calibrate accuracy of loss estimation models, incorporating the full range of physical and societal impacts and losses
- National models for seismic hazards, building and lifeline inventories, and exposed populations with application to other natural and man-made hazards
- Improved damage and fragility models for buildings (structural and nonstructural) and lifelines

- Improved indirect economic loss estimation models
- Development of system-level simulation and loss assessment tools that address lifeline interdependency issues (also addressed in Tasks 12 and 15)

The use of "default" data or simplifications of data contribute to the uncertainty in earthquake risk estimation. The aggregation of building inventory data to the census tract or census block level and the use of model building types with simplified fragility curves and model-driven databases may misrepresent the actual characteristics of building inventories. Differences between Assessor's databases and the HAZUS database tend to underestimate nonresidential exposure in large urban counties and overestimate exposure (square footage) in smaller, less urban counties (Seligson, 2007). Improvements to HAZUS data management tools—such as the HAZUS Comprehensive Data Management System[16] (CDMS)—help to address and reduce some of these uncertainties by allowing communities to import higher resolution data.

2. Promote a "living" community risk model. Because our communities are changing all of the time, community resilience is a dynamic concept. Optimum decision-making, at all levels of society, depends on the availability of current, up-to-date information. Risk assessment needs to be made real to the community by being open and accessible to users. The ability to define what the current acceptable level of disruption is at the local community level requires flexibility to incorporate:

- Local inventory data from various sources.
- New information and data (i.e., new attenuation models, building fragility curves, demographics, lifeline performance models, network interdependencies, indirect economic losses).
- New software or improve upon existing software, such as front-end and back-end software modules (e.g., programs that can address lifeline network disruptions and network interdependencies).

In addition to the basic risk metrics already available (e.g., direct/indirect economic loss, causalities, debris), the development of new analysis techniques or new metrics that may be specific to an individual or to the needs of a community should be supported.

Establishing community risk models for Earthquake-Resilient Community and Regional Demonstration Projects would be one means of

---

[16] See www.fema.gov/plan/prevent/hazus/hz_cdms2.shtm.

showcasing how risk assessment can be used to inform risk reduction activities.

Existing Knowledge and Current Capabilities

The ability to compare risk across states and regions is critical to the management of NEHRP. Loss assessment tools provide uniform engineering-based approaches to measure damages and economic impacts from earthquakes. Many of these models are contained in commercial software packages that have been developed by firms specializing in the development and marketing of proprietary models to end users (e.g., the insurance industry), and include those developed by AIR Worldwide, EQECAT, Risk Management Solutions, and URS. In addition to these proprietary earthquake loss estimation programs, there are currently two publicly available loss estimation or risk assessment programs—FEMA's HAZUS, and the Mid-America Earthquake Center's MAEviz program:

- FEMA developed HAZUS in cooperation with the National Institute of Building Sciences (NIBS), and by 2010 had released two generations of software. The first release, HAZUS-99, only addressed earthquakes, whereas the HAZUS-MH releases address flood and wind as well as earthquakes.[17]
- MAEviz is a joint effort between the Mid-America Earthquake Center (MAE) and the National Center for Supercomputer Applications (NCSA) to develop open-source seismic risk assessment software based on a Consequence-based Risk Management methodology. Open-source architecture helps to reduce the time lag between discovery by researchers and implementation by end users. New research findings, software, improved methodologies, and data can be added to the system using a plug-in system. As a result, MAEviz is constantly changing and evolving with daily builds posted on the web.[18]

Another model, an international open-source code program called the Global Earthquake Model, is currently under development and is scheduled for release by the end of 2013.[19]

Uses of Risk Assessment and Loss Estimation Modeling.

Risk assessment and loss estimation modeling has been successfully used at both national and community scales to promote awareness of

---

[17] See www.fema.gov/plan/prevent/hazus/index.shtm.

[18] See mae.cee.uiuc.edu/software_and_tools/maeviz.html.

[19] See www.globalquakemodel.org/.

earthquake risks. USGS Circular 1188 (USGS, 1999) multiplied earthquake hazard (10 percent chance of exceedance in 50 years) by population size to create a risk factor that was used to identify the number of urban seismic stations needed as part of ANSS (see Task 2). FEMA 366 (FEMA, 2008) provides a national estimate of the long-term average annual earthquake loss to the general building stock (see Box 1.1) based on the HAZUS methodology. The current AEL for the United States, based on the 2000 Census, is $5.3 billion (2005$). As seen in Table 3.2, 43 metropolitan areas—led by Los Angeles and San Francisco—account for the majority (82 percent) of the earthquake risk in the United States. Outside of California, at risk communities including Seattle, WA, Portland, OR, Salt Lake City, UT, and Memphis, TN, show that earthquakes are not just a California problem.

Loss estimates can also be used to gauge the effectiveness of various mitigation strategies such as building retrofitting or the transfer of risk through the sale of property or the purchase of earthquake insurance. FEMA (1997b), for example, estimated that the direct economic losses (building and contents damage, and income losses) in an event similar to the 1994 Northridge earthquake would have been reduced by 40 percent ($16.6 billion compared to $27.9 billion) if all buildings had been built to current high seismic design standards prior to the earthquake. Had no seismic standards been in place, losses were estimated to have been 60 percent higher than those for the baseline 1994 scenario ($45 billion versus $27.9 billion). A 2001 FEMA report, based on the HAZUS-99 earthquake loss estimation methodology, examined the impact of seismic rehabilitation in reducing the economic and social losses from magnitude-7 earthquakes on the Newport Englewood Fault in southern California and the Hayward Fault in northern California (Feinstein, 2001). In both cases, the HAZUS modeling indicated that a comprehensive rehabilitation program could reduce building and contents damage losses by more than 25 percent and business interruption losses by more than 60 percent. These types of retrospective loss avoidance studies show how future losses can be recognized and avoided through simulation modeling and proactive community mitigation programs.

Loss estimates are affected by uncertainty—uncertainty in estimating the likelihood and intensity of strong ground motion, uncertainty in actual community building and infrastructure inventories, uncertainty concerning the levels of damage to the built environment, and uncertainty in the social and economic losses associated with the predicted damage. These uncertainties also impact estimates of financial risk and the premiums that insurers charge for earthquake insurance (NRC, 2006b). The high cost of earthquake insurance, resulting in part from the uncertainty in estimates of seismic risk, limits the amount of earthquake insurance purchased. Analyses conducted as part of the 140th anniversary of the 1868

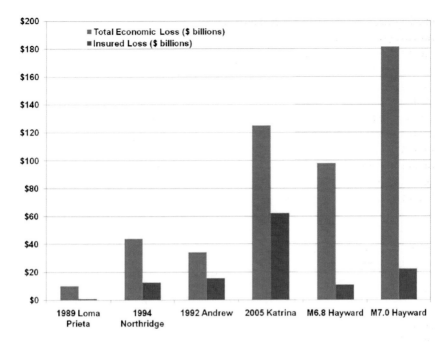

FIGURE 3.13 Comparison of insured and economic losses from recent U.S. natural disasters (in 2008$). Insured losses for Hurricanes Andrew and Katrina include National Flood Insurance Program (NFIP) policies. SOURCE: Risk Management Solutions; Zoback and Grossi (2010).

Hayward, CA, earthquake indicate that only 6 to 10 percent of total residential losses and 15 to 20 percent of commercial losses would be covered by insurance following a repeat of the magnitude-6.8 to 7.0 earthquake (RMS, 2008). In contrast, approximately 53 percent of the economic losses to homes and businesses following hurricane Katrina were covered by insurance, including payouts from the National Flood Insurance Program (Figure 3.13).

Enabling Requirements

To continue the progress already made in community earthquake risk assessment, continued NEHRP-funding for the development of nationally consistent datasets—such as the National and Urban Seismic Hazard Maps discussed in Tasks 4 and 6, and improved fragility curves for model building types that account for regional differences in construction practices, code levels, and structural condition—is essential. Support for an open cyber environment that supports the continual update and improvement of risk assessment software, and the continued development of new basic physical models (e.g., fire following earthquake) is also necessary.

In addition to these types of national-scale products, the NEHRP agencies also work at the community level providing expertise and data for knowledgeable risk management activities. FEMA has distributed HAZUS throughout the United States and has been instrumental in establishing local HAZUS User Groups (HUGs) that provide local GIS support and expertise to communities. USGS works with state and local agencies to improve urban earthquake hazard maps and provides guidelines and procedures for collecting site condition information at the community level. Mapping local geology in three dimensions and incorporating more detailed grids into maps of site response, liquefaction, and landslide potential increase the granularity of these data, which then improve the resolution of the community earthquake risk assessments.

Implementation Issues

Open- Versus Closed-Source Software

In addition to informing risk assessment and mitigation activities, the HAZUS loss estimation software—which is "closed-source" software (i.e., the source code is not available to the community)—is also used as a decision support tool for emergency management (e.g., for requests for Presidential Disaster declarations and the development of State Mitigation Grants). Although a standard source code is necessary for consistent national decision-making, an open version where developers can test new data and develop algorithms should be supported as well. Increasingly, the scientific and engineering community is making use of "open-source" software, to create a cyber environment where new data, concepts, and applications can be developed and tested. Community model environments like MAEviz and the Open System for Earthquake Engineering Simulation (OpenSees)[20] provide a software framework for regional- or community-based scenario development and for simulating impacts and the seismic response of structural and geotechnical systems. Linking the open-source environment to the risk and loss assessment development process would enable faster application and implementation of research results. Once these new models and concepts have been appropriately vetted, they can be incorporated into a more standardized platform for use by the NEHRP agencies.

---

[20] See opensees.berkeley.edu/index.php.

Community Adoption/Implementation

It would be useful for the NEHRP agencies to implement a series of coordinated activities tied to the Earthquake-Resilient Community and Regional Demonstration Projects, discussed in Task 18. These activities would provide the basic data and mapping (through the USGS Urban Hazard Mapping projects) and building and infrastructure inventories (through FEMA-supported HAZUS activities) to support community risk management activities.

Confidentiality Issues

Many stakeholders, especially those in areas of critical infrastructure, are reluctant or, because of provisions in the Homeland Security Act of 2002, are unable to release inventory information beyond their organizations. These restrictions impact the ability of communities to recognize and plan for service disruptions during disasters. Public-private partnerships should be encouraged, where individual utilities and lifeline organizations conduct their own internal risk assessments—using standardized methodologies and earthquake scenarios—and then share the results with their counterparts and other stakeholders to address inter-utility interdependencies and community impacts from the loss of utility service. These types of partnerships would permit more informed disaster planning within the community.

TASK 8: POST-EARTHQUAKE SOCIAL SCIENCE RESPONSE AND RECOVERY RESEARCH

Summarized most recently in NRC (2006a), early to more recent social science research under NEHRP has highlighted, on the one hand, major obstacles to achieving anything more than modest levels of pre-disaster mitigation and preparedness practices at household, organizational, community, and regional levels, and on the other hand, the often extraordinary resilience at all these levels of human response during and after actual earthquakes and other events (Kreps and Drabek, 1996; Kreps, 2001; Drabek, 2010). In so doing, research over decades has contradicted misconceptions that during a disaster panic will be widespread, those expected to respond will abandon their roles, social institutions will break down, and anti-social behaviors will become rampant. The more important research questions have become how and why communities and regions are able to leverage expected (and perhaps planned) and improvised emergency response and recovery activities in both the public and private sectors.

There are major demands and considerable public pressure in the immediate post-disaster environment to return to normalcy as quickly as possible (Kreps, 2001; Tierney et al., 2001; Tierney, 2007; Johnson, 2009). That is why social science studies of expected and improvised emergency response activities under NEHRP's legislative research mandate continue to be important. Yet, social science research has suggested also that the post-disaster environment provides one of the most opportune times for disaster recovery activities to support hazard mitigation—to rebuild stronger, change land-use patterns, and reduce development in hazardous areas, and also to reshape those negative social, political, and economic conditions that existed pre-event (NHC, 2006; NRC, 2006a; Olshansky et al., 2006). Thus, just like emergency response activities, disaster recovery activities need to be prepared to the extent possible, and then executed appropriately to reduce future risks. The NEHRP agencies, most notably FEMA, have responsibility for many of the federal programs that provide funding to communities and regions for emergency response and recovery. Thus, the social science research under NEHRP proposed here aims to ensure that its related mission to enhance community and regional resilience can be more fully realized.

Proposed Actions

Fundamental social science studies are needed of post-disaster practices that increase the resilience of communities and regions following large-scale earthquakes and other major disasters (see also Tasks 10 and 11). Such studies will document and model the mix of expected (and perhaps planned) and improvised emergency response and recovery activities and outcomes at community and regional levels, as they are supported in varying degrees by the federal government. The primary research targets on emergency response and recovery activities are governmental, medical, and educational organizations, social services agencies, public utilities, and industrial and commercial organizations. The disaster demands to which these entities must respond include mobilizing emergency personnel and resources, evacuation and other types of protective action, search and rescue, care of victims, damage assessment, restoration of lifelines and basic services, reconstruction of the built environment, and maintaining continuity of the economy and government. The studies we propose will contribute to both NEHRP's legislative research mandate and its related mission to enhance community and regional resilience (see also Task 18).

Existing Knowledge and Current Capabilities

While the clear majority of pre- and post-disaster practices at community and regional levels are expected and sometimes planned, improvisation is an absolutely essential complement of pre-determined activities. Heretofore, social science studies of *emergency response practices* have been given primary attention under NEHRP and, to some extent, *the pre-disaster preparedness practices related to them*. These studies have documented the mix of expected and improvised activities of emergency management personnel, the public and private organizations of which they are members, and the multi-organizational networks within which these individual and organizational activities are nested (e.g., Kreps and Bosworth, 2006; NRC, 2006a; Mendonca, 2007).

Little research has focused on pre- and post-disaster recovery practices (expected or improvised) in either the public or private sectors (NRC 2006a). However, the outcomes of these practices are increasingly being given focused attention by social scientists (e.g., NRC, 2006a; Rose, 2007; Alesch et al., 2009; Olshansky and Chang, 2009; Zhang and Peacock, 2010). Accordingly, the proposed research has three primary aims: *first*, to build on existing knowledge of emergency response and related preparedness practices; *second*, to expand knowledge about disaster recovery and related preparedness practices; and *third*, to develop models and decision support tools that are increasingly grounded in social science knowledge about disaster response and recovery. The use of such models and tools, we believe, will enhance community and regional resilience before, during, and after earthquakes and other disasters.

The emergency response improvisations that have been documented systematically for a broad range of disasters include the following:

- *At the individual level,* the spontaneous adoption of important post-disaster roles by individuals who, based on their pre-disaster positions, would not be expected to do so. In effect, such individuals rise to the occasion because they happen to be in the right place and at the right time when there is a compelling demand for action and often leadership.
- *At the individual level,* the spontaneous development of new as opposed to pre-existing relationships among individuals performing important post-disaster roles. New relationships are forged because they facilitate the performance of roles by either or both partners in the relationship.
- *At the individual level,* the unconventional performance of post-disaster roles regardless of whether they are pre-determined or spontaneously adopted, and regardless of whether they are facilitated by pre-existing or new relationships. The improvisations can range from procedural or equipment changes related to how the roles are enacted, to changes in the usual locations of the role enactments, to taking on activities that are not authorized, to the issuing of orders to others over whom there is no pre-existing authority, and to the commandeering of supplies and equipment without prior approval.

- *At the individual level,* the primary reasons for the above improvisations are human and material resource needs, operational issues, time pressures to get things done, and frequently mixes among these kinds of problems and opportunities at either intra- or inter-organizational levels of response.
- *At the organizational level,* the timing and location of core activities may be changed, human and material resources may be reconfigured, some core tasks may be suspended while others are expanded or newly created, and, in some cases, relatively complete short-term reorganizations of pre-disaster routines may occur.
- *At the multi-organizational response network level,* the most frequent types of improvisation relate to unconventional exchanges of human and material resources, newly coordinated activities, or unconventional exchanges of resources in association with newly coordinated activities. More elaborate arrangements that involve changes of authority patterns or one or more organizations being absorbed within more inclusive entities are not common. But such a result is not inevitable because completely new organizations of various forms and sizes have sometimes arisen during large-scale disasters and proven to be consequential.

The individual, organizational, and multi-organizational improvisations that have been documented by previous and ongoing social science studies relate primary to *immediate post-disaster demands for emergency services as opposed to short- and longer-term demands for reconstruction and recovery.* An important difference between emergency response and recovery is that the key players in the former (e.g., police, fire, emergency medical services, public utilities, local emergency management offices) and the latter (e.g., community development agencies, land-use boards, real estate companies, banks, insurance companies, local businesses) generally do not interact routinely and have different organizational cultures. However, the data collection tools that have been used to codify the mix of expected and improvised activities by emergency response personnel in the public sector can be applied across the board. But if data collection is to become more comprehensive, standardized research protocols on expected and improvised activities within and between the public and private sectors must be developed and new arrangements for data archiving, data management, and data sharing must be created.

Enabling Requirements

Social scientists studying earthquakes and other hazards have used a myriad of research methods. They have employed both quantitative and qualitative data collection strategies. They have conducted pre-, trans-, and post-event field studies of individuals, households, groups, and organization. These studies have relied on open-ended to highly structured surveys and face-to-face interviews. They have used public access data such as census materials and other historical records from public and private sources to document community and regional vulnerabilities to earthquakes and other hazards. They have employed spatial-temporal data and related statistical models to document these vulnerabilities as well. They have engaged in archival studies of previous events when data from the original studies have been stored and made accessible. They have run disaster simulations and gaming experiments in laboratory and field settings. The social science research methods heretofore been used have been enabled by both "off the shelf" and cutting-edge technologies (NRC, 2006a, 2007).

Three key enabling requirements relate directly to the post-disaster response and recovery research proposed above: *standardized data collection, improved data management, and sustained model building.* Meeting these requirements will support the development and use of management support tools by those actively engaged in emergency response and recovery at local and regional levels.

• *Standardized Data Collection:* Post-disaster data collection by social scientists historically has been undertaken under very difficult conditions (see also Task 9). The timing and location of field observations have been heavily constrained by the circumstances of the events themselves as have the possibilities to make audio and video recordings of response activities. There have been special constraints and difficulties in sampling of and collecting data on emergency response personnel, their organizations, and social networks of responding organizations within the public and private sectors. Unobtrusive data such as meeting minutes, formal action statements, communications logs, memoranda of understanding, telephone messages, and email exchanges are difficult and sometimes impossible to obtain, and so on. The proposed pre-selection of community and regional demonstration projects (see also Task 18) and collection of data on a cross-section of communities prior to and following disasters (see also Task 11) are therefore very important for reducing the inherent ad hoc quality of most previous post-disaster studies. The cooperation of key organizations in the public and private sectors can be secured and obviously will become essential if or when actual events occur. With that cooperation, sample frames of those engaged in emergency response and recovery activities can be pre-determined to a much larger extent than has been possible in the past. Standardized data protocols on expected and improvised activities and their determinants can be developed and made ready on a standby basis. Methods of data storage and agreements on data sharing can be set up before rather after the fact. Previous attempts to standardize social science data on earthquakes and other hazards have been intermittent and not well-coordinated among respective individual researchers or teams working on the same or related topics. But the potential for standardization in future studies is enormous. Simply put, social scientists now know what to look for in studies of post-disaster response and recovery. State-of-the-art computing and communications technologies can be used to implement data protocols more efficiently and effectively than in the past.

  • *Improved Data Management:* Once collected, data archiving and dissemination become very important functions that also are enabled by computing and communications technologies (see also Tasks 4, 6, and 11). Advances in data standardization will compel the application of technical tools to support the management of archives and mining data from them. Much can be learned about these functions from ongoing research and development activities in the physical and life sciences, engineering, and interdisciplinary work in computational science (e.g., software solutions and professional services that support extraction of data, visual imaging, and web browsing). The growing technical capabilities and required bandwidth for data transmission through the Internet certainly will facilitate data sharing efforts within all natural science, social science, and engineering fields (see NRC, 2007, for description of benefits for studies of earthquakes and other hazards). Technologies that enable data archiving, data mining, and data dissemination must be augmented in the immediate future by formal management of data sharing. The "rules of the game" on data sharing are not nearly as clear and agreed upon as those related to the control of data by original investigators. Formal data control values and norms (i.e., those related to standards of validity and reliability, proprietary access and intellectual property, human subject protection, confidentiality of information, and anonymity of sources) must translate as formal "terms of use" in the sharing of data between original researchers and secondary data analysts. It is essential for those studying disasters to consider formally the management of data and to promulgate formal standards for data sharing before rather than after data standardization and building archives gain momentum.

  • *Sustained Model Building:* Modeling is the sine qua non of science (see also Tasks 6, 7, 10, 12, and 17). Its goal is to help researchers and practitioners alike to better understand how the world works. Technological advances in computing have enabled the development of complex models that can be used to describe and explain phenomena in both physical and social systems, from the smallest to most inclusive imaginable. And these advances have contributed greatly to the development of interdisciplinary research. An important use of computing in the natural sciences, social sciences, and engineering remains statistical models of existing data. These statistical models range from relatively simple to highly complex configurations of variables. They are being used increasingly to create structural models of post-disaster response and recovery practices and outcomes. And over time, the expanding computing capacity has enabled the development of decision-making models as well. Decision-making models often rely on simulations and other forms of field or laboratory experimentation that place subjects (e.g., emergency response and recovery practitioners) in hypothetical situations (e.g., disaster circumstances) to see how decisions are made and actions taken (see also Task 6). It is important to emphasize that these decision-making models are theoretically driven, and their power is enhanced to the extent they are empirically based (NRC, 2006a). In combination, structural and decision-making models can serve as a key foundation for development and use of preparedness and training tools at community and regional levels (see also Task 18).

Implementation Issues

Three central implementation issues, and their possible resolution, merit serious consideration: lack of predictability about when large-scale earthquakes and other major disasters will occur; current lack of standardized research protocols on post-disaster response and recovery activities and related pre-disaster preparedness practices; and lack of standby research facilities and capabilities to implement standardized research protocols and manage data resulting from their use.

•   *Predictability of large-scale earthquakes and other major disasters:* At community and regional levels, disasters are low probability events and, as such, very difficult to predict. Accordingly, research sites for post-disaster studies are largely ad hoc, there are major difficulties in mobilizing field research teams quickly, and emergency contexts present serious difficulties for data collection. Despite these historical research barriers, social scientists have been able to collect a wide range of ephemeral data on emergency response and recovery activities. The pre-selection of pilot communities and regions during the next 5-20 years of NEHRP's Strategic Plan (see Tasks 11 and 18) will facilitate social science research greatly, first, because research plans can be developed, second, because by-ins by local and regional officials in the public and private sectors will be more likely, and third, because the likelihood of one or more events occurring during the next 5-20 years in the pilot research sites will be more likely (NRC, 2006a).

•   *Lack of standardized data collection protocols:* Much of the groundwork has already been established, thus the potential for highly structured research designs and replicable datasets across multiple disasters can now be realized. The key requirement is to have standardized data collection protocols on emergency response and recovery activities already in place before specific events occur (NRC, 2006a). To that end, we suggest that NEHRP agencies fund as soon as possible, under the auspices of the National Science Foundation, a specific initiative on the development of these research protocols. The competition should attract existing or new research teams interested in related methodological issues. The budget to fund 2-4 projects during the next 2 years, excluding the cost of actual data collection, should be on the order of $1.5 million.

   • *Lack of standby research facilities and capabilities to implement research protocols and address related data management issues:* The development of standardized research protocols needs to be matched by the existence of standby research capabilities and facilities to collect, manage and disseminate resulting data. Existing university-based social science research centers could serve this purpose in the near term through newly designated funding. But ultimately, as recommended in NRC (2006a), a National Center for Social Science Research on Earthquakes and Other Disasters is needed. Such a center would include a distributed consortium of investigators and research units nationally and internationally. Similarly to Network for Earthquake Engineering Simulation (NEES), it would take advantage of telecommunications technology to link spatially distributed data repositories, facilities, and researchers. It would provide an institutionalized, integrative forum for social science research on hazards and disasters, much as the Southern California Earthquake Center (SCEC) does for the earthquake earth sciences community. We suggest that the NEHRP agencies provide funding for the new social science center, under the auspices of the National Science Foundation, for the next 5 years, at an annual budget of $2 million per year. Such funding would be consistent with previous NSF funding of earthquake research centers.

TASK 9: POST-EARTHQUAKE INFORMATION MANAGEMENT

   Although catastrophic earthquakes are rare, damaging earthquakes occur more frequently. Capturing, distilling, and disseminating lessons about the geological, structural, institutional, and socioeconomic impacts of earthquakes, as well as the responses post-disaster, are critical requirements for advancing knowledge and more effectively reducing earthquake losses. The 2008 NEHRP Strategic Plan for 2006-2010 identifies the creation and maintenance of a repository of important post-earthquake reconnaissance data as a strategic priority to improve understanding of earthquake processes and impacts (NEHRP, 2007). This task aims to ensure that

NEHRP's activities would be more effective in the post-disaster period by improving post-earthquake information acquisition and management.

Proposed Actions

This task proposes to construct and maintain a national post-earthquake information management system to capture, distill, and disseminate lessons from damaging earthquakes. The system, in itself, will be a significant engineering effort, and it will also require sustained multi-year funding to implement and maintain in order to cost-effectively preserve data over time, so it will still be accessible and usable for future infrastructure design, and for mitigation and disaster management efforts. It will help ensure that NEHRP's mission—to develop, disseminate and promote knowledge, tools, and practices for earthquake risk reduction in the pre-disaster environment—can also be successful in the post-disaster environment.

Existing Knowledge and Current Capabilities

It has long been recognized that any national effort to reduce economic losses and social disruption resulting from severe natural disasters requires a mechanism to capture and preserve engineering, scientific, and social performance data in a comprehensive and coherent system that will contribute to our learning from each disaster event that occurs (EERI, 2003a). Such a resource would play a vital role in efforts to enhance infrastructure design and to optimize mitigation, disaster planning, and response and recovery efforts. Despite this recognition, no mechanism is currently in place across the United States to ensure that necessary data are systematically collected and archived for future use. Further, those data that are gathered often are lost relatively soon after they have been collected, instead of being organized and maintained to enable study, analysis, and comparison with subsequent severe natural disasters that may not occur for many years or even decades (NRC, 2006a).

There are many agencies and professional organizations that currently support or are involved in post-disaster information acquisition and management. They include NSF's funding of EERI's Learning from Earthquakes[21] and Geotechnical for Extreme Events Reconnaissance Association;[22] both are working on more systematic approaches to conducting the NSF-sponsored reconnaissance efforts of the effects of extreme events. USGS is also very active in post-disaster reconnaissance, both in

---

[21] See www.eeri.org.
[22] See www.geerassociation.org.

the United States and internationally, and it has developed a plan to help coordinate NEHRP Post-Earthquake Investigations (USGS, 2007).

Recently, FEMA funded some initial scoping of the requirements for such a system under the auspices of the Multihazard Mitigation Council's American Lifelines Alliance (ALA). The goal of the ALA effort was to identify both infrastructure requirements (e.g., data system architecture, technological needs and issues), and implementation requirements (e.g., facilities, expertise, policies, and funding) for a Post-earthquake Information Management System (PIMS). A PIMS would provide users with the ability to query data in an intuitive and interactive manner to investigate the past performance of the built environment during earthquakes.

The ALA held a Workshop on Unified Data Collection in Washington, DC, on October 11-12, 2006 (NIBS, 2007). The workshop served as a forum for open and candid discussion of common needs of the utility and transportation systems (lifelines) community and possible opportunities for cooperation and collaboration in addressing those needs. The findings from the workshop contributed to the identification of "Improve post-earthquake information acquisition and management" as an objective and "Develop a national post-earthquake information management system" as a strategic priority in the 2008 NEHRP Strategic Plan (NIST, 2008). The workshop participants recognized that an integrated PIMS needed to include all aspects of the built environment and could potentially be expanded in scope to address all types of natural hazards.

In December 2007, the ALA, with funding from FEMA, tasked a team of researchers from the University of Illinois to conduct a 10-month scoping study to assess user needs and system requirements, challenges, and system-level issues for implementing a PIMS, and the design strategy needed to overcome the challenges and satisfy user needs (PIMS Project Team, 2008). As a follow-on project, researchers at the University of Illinois utilized "wiki" technology to collect summaries of information needs and applications, so that anyone can review and edit existing summaries or add a new summary. Funding and implementation of the ALA project did not occur, nor has integration with GEER, EERI, USGS and other more recent efforts.

Enabling Requirements

Building a more earthquake-resilient nation will require better systems to capture, distill, and disseminate lessons from damaging earthquakes. Development of a PIMS would be a significant engineering effort, and will require sustained multi-year funding to implement a system capable of cost-effectively preserving data for 50 to 100 years. There are both user needs and system requirements/issues for a PIMS (PIMS Project Team, 2008):

- User Interfaces: the interfaces that users would prefer to use for the discovery and retrieval of PIMS data.
- Information Needs: the types of information that users want to obtain from PIMS, including a range of general information, hazard data, building data, bridge data, lifeline data, critical structures data, historical data, loss/socioeconomic data, and pre-event inventory data.
- Data Access, Privacy, and Security Issues: a PIMS would need to respect data privacy, consistent with state and federal laws, by removing personal information from data, creating aggregated datasets, and restricting access to certain types of data.
- Direct Ingestion of Data: there would need to be the capability to directly upload data into PIMS.
- Harvesting and Exchange of Data: a PIMS would need to be able to harvest and exchange data with a variety of existing electronic databases.

In addition to serving the direct needs of users and other stakeholders, a PIMS must address their implicit assumptions about how the system's scope should align with their goals. A PIMS also must address issues related to the cultural, political, technological, and organizational context in which it will operate. System requirements and system-level issues have been identified that relate to data collection, organization, and storage; data curation and quality assurance; information presentation, discovery, and retrieval; privacy and security; and long-term data preservation.

Implementation Issues

PIMS is similar to NSF's national environmental observatory efforts, and the overall timescale from project initiation to mature operational capability is 5 to 10 years; however, it could occur in two phases (PIMS Project Team, 2008):

- An initial PIMS capability (PIMS Phase 1) could be accomplished in 2 years and would include development of an initial PIMS capable of harvesting data from a few key sources, basic ingestion and archiving capability for hazards events in the near future, and a simple interface to provide for data discovery and retrieval.
- Phase 2 would take from 5 to 10 years and would involve development of a more advanced, "full-function" PIMS capable of harvesting data from a wide variety of sources, providing advanced tools for ingesting and archiving and offering sophisticated user interfaces for data discovery

and retrieval. Phase 2 will involve about seven to nine pilot projects that would have both a development phase and an implementation phase. Operations costs would continue beyond the development period of Phase 2.

TASK 10: SOCIOECONOMIC RESEARCH ON HAZARD MITIGATION AND RECOVERY

Social science research complements research in other fields of earthquake resilience. For example:

- Hazard loss estimation, including its extension to macroeconomics, helps us establish an understanding of the severity of the earthquake problem to society.
- Psychology helps us understand how people perceive the earthquake threat and the need to address it.
- Decision science and behavioral economics assess the motivations and prudence of individual decision-making with respect to this threat.
- Organizational behavior analyzes group decisions in the realms of business, government, and nonprofit organizations.
- Sociology emphasizes how individuals and groups interact under stress in the aftermath of an event.
- Economics and finance help provide guidance on the allocation of funding to projects and policies, including insurance.
- Planning examines how the built environment can be modified by structural and nonstructural approaches in a cohesive fashion.
- Political science indicates how research, resource availabilities, and debate are translated into laws and regulations by several levels of government.

In the post-disaster environment, governments—particularly local governments—face considerable public pressure to provide a quick return to normalcy. Yet, research has consistently shown that the post-disaster environment provides one of the most opportune times for mitigation—to rebuild stronger, change land-use patterns, and reduce development in hazardous areas, and also reshape social, political, and economic pre-existing conditions—and thereby helps to break the repetitive loss cycle (Berke et al., 1993; Schwab, 1998; Mileti, 1999; NHC, 2006; NRC, 2006a). Long-term recovery needs time to be accomplished thoughtfully and to allow for proper deliberation and public discourse on how to achieve risk reduction and betterments. The NEHRP agencies, most notably FEMA, have responsibility for many of the federal programs that provide funding to communities to recover from an earthquake or other damaging disasters. This action aims to ensure that NEHRP's mission can be more effective in the post-disaster period by promoting support to increase the earthquake resiliency of impacted communities, including mitigation ahead of the next disaster.

Proposed Actions

Basic and applied research in the social sciences, as well as such related fields as business and planning, is needed to evaluate mitigation and recovery (both short-term business and household continuity and long-term economic and community viability). Such studies would examine individual and organizational motivations to promote resilience, the feasibility and cost of resilience actions, and the removal of barriers to successful implementation. They should focus on the appropriate roles for both the private and public sectors, and look toward partnerships that avoid one sector undercutting the appropriate role of the other. Improved data and models are needed at the basic research level that will promote a sounder basis for policy prescriptions. Key hypotheses should be tested to give the models much needed behavioral content. Benefit-cost and other evaluative studies of pre-disaster mitigation and post-disaster resilience are encouraged to improve the management of our nation's resources.

Task 8 complements this section by addressing emergency response and related short-term recovery. Task 11 recommends the creation of an Observatory Network that would help promote these goals in part, especially with respect to on-going data collection and analysis. However, this section covers a broader range of activities. Also, continued sponsorship of individual researchers and established and new research centers is encouraged to promote innovation, practical tools, and policy advice on resilience to earthquakes

Existing Knowledge and Current Capabilities

Pre-Disaster Loss Prevention

On the mitigation side, research is well advanced in the social sciences. Studies have found that FEMA hazard mitigation grants yield a benefit-cost ratio of 4 to 1 overall, and 1.5 to 1 for earthquakes (MMC, 2005; Rose et al., 2007). The reason that the ratio for earthquakes is lower than for other threats is that earthquake mitigation has focused much more on life saving than on property damage, and because there are more easily implemented mitigation actions (e.g., buy-outs of homes in flood plains) for these other threats. This study of mitigation projects, however, was highly weighted toward government initiatives, and more research is needed for private-sector efforts. It is important to overcome the temptation to say that the marketplace will guide business decisions-makers to the optimal level of mitigation because of the profit motive. The existence of externalities of individual decisions, i.e., impacts of one decision-maker that affects others in some positive or negative way, is prevalent in this area. An excellent example of the extent of this issue is the work on "contagion effects" by Heal and Kunreuther (2007), which points out the limitations of one person undertaking protective measures if neighbors do not, such as in the case of the fire threat following earthquakes.

The Multiharzard Mitigation Council (MMC) study only scratched the surface in understanding the effectiveness of individual process grants and broader mitigation strategies, as well as general resilience, capacity-building grants. The former refer to funding for activities such as earthquake mapping and monitoring systems. The latter refer to broader block grants such as Project Impact. These are difficult to assess, because they are fewer in number and it is difficult to measure their effectiveness (Rose et al., 2007).

One of the greatest research needs in the pre-event phase is still, after many years, why individuals and businesses fail to make rational decisions about self-protection and insurance (Ehrlich and Becker, 1972; Jackson, 2005). Some excellent studies have identified limitations of the decision-making process under the category of "bounded rationality" (Gigerenzer, 2004). This includes classic work by Kunreuther et al. (1978) on understanding the failure of people to purchase adequate amounts of earthquake insurance. More research is needed on how to counter this problem, including overcoming myopia and other perception issues, dealing with moral hazard, and determining how government policy can inspire individual motivations as opposed to undercutting them. An excellent start in the general areas of hazards and terrorism has been provided by Smith et al. (2008) and Kunreuther (2007).

The research on community resilience has made substantial progress in just a few years (e.g., Norris et al., 2008). This research still faces challenges and can also yield many valuable spin-offs to researchers pursuing interdisciplinary and comprehensive approaches to resilience.

Predictive models of individual and community resilience are also valuable. Some initial attempts, analogous to Cutter's design of a vulnerability index (Cutter et al., 2003), are being developed (Schmidtlein et al., 2008; CARRI, 2010; Cutter et al., 2010; Sherrieb et al., 2010).

Disaster Recovery and Reconstruction

Most post-disaster resilience activities are intended to reduce business interruption. Nearly all property damage occurs during the ground shaking, but business interruption begins at that point and continues until the recovery is complete. The analysis is complicated by the fact that business interruption has extensive behavioral and policy connotations. For example, it is a factor of the length of time it takes to recover, which is not constant but highly dependent on the mix of individual motivations and government policies.

Operational metrics that can be applied to measure post-disaster resilience have been developed and applied effectively (Chang and Shinozuka, 2004; Haimes, 2009). Studies have examined the relevant contribution of various types of resilience in reducing losses from natural disasters and terrorism (e.g., Rose and Liao, 2005; Rose et al., 2007). However, only a few studies have actually evaluated the costs of these various resilience strategies (e.g., Vugrin et al., 2009). A priori, one would expect these strategies to be much less costly than pre-disaster mitigation. Conservation of inputs that have become even scarcer usually pays for itself, the cost of inventories is merely their carrying cost, and production recapture at a later date requires only the payment of overtime for workers.

Enabling Requirements—Needed Studies

The following areas of research are needed to better understand post-disaster resilience actions individually and as a group:

- Inventory of the many actions that can be undertaken to implement resilience after an event. This pertains to three levels (Rose, 2009)—microeconomic (individual household, business, or government entity); mesoeconomic (an entire industry or market); and macroeconomic (the entire economy, including interactions between decision-makers and institutions).
- Assessment of the efficacy of actions that can be taken to enhance this resilience prior to an event (e.g., the building up of inventories, emergency drills) or after the event (e.g., relocating businesses quickly and matching customers who lost their suppliers with suppliers who lost their customers). The extent to which business interruption losses can be reduced. Studies by Tierney (1994), Rose et al. (2007), and Kajitani and Tatano (2007), for example, indicate that the potential for reducing losses is great for selected approaches to resilience. However, many types of resilient actions have not yet been assessed.
- Estimation of the costs of implementing resilience. The studies just noted also give some indication that many post-disaster resilient activities are relatively inexpensive. Conservation pays for itself, inventories only incur carrying costs, and production rescheduling simply requires the

payment of overtime for employees. Still more studies are needed to cover the entire range of resilience alternatives.

• Evaluation of the emerging business continuity industry. An increasing number of private firms offer disaster recovery services (Rose and Szelazek, 2010). This professionalization of recovery is likely to improve the efficiency of recovery and reduce the need for government assistance. Still, the broader implications of this industry need to be ascertained with respect to adherence to professional standards, potential market power, price gouging, and affordability to small business.

• Organizational response. Research by Comfort (1999) provided a valuable foundation in terms of nonlinear adaptive systems. Such research captures the evolving nature of the institutional decision-making process, including learning and feedback effects. More case studies are needed.

• Identification of obstacles to implementation. Most studies of resilience to date focus on an ideal context in which a resilient action is implemented. Various types of market failures, transaction costs, regulatory restrictions, and limited foresight need to be understood first before research on how they might be overcome can proceed (e.g., Boettke et al., 2007; Godschalk et al., 2009).

• Identification of best-practice examples. There are notable examples of successful resilience, as in the use of backup electricity generators in the aftermath of the Northridge earthquake and business relocation following 9/11. Research analyzes underlying issues; however, practitioners are more likely to be won over by real-world successes. (e.g., Tierney, 1997; Rose and Wein, 2009).

• Design of remedial policies. This includes innovative approaches, such as the use of vouchers and other incentive-based instruments to promote resilient actions. Especially critical is research that identifies areas in which the public and private sector can work in harmony to achieve resilience, rather than at cross purposes. Many have pointed to government bailouts as providing a disincentive for both mitigation and resilience, although it would help to try to measure the extent to which this takes place (e.g., Smith et al., 2008).

• Characterization of infrastructure network vulnerability and resilience. Resumption of infrastructure services is one of the primary needs of recovery. Network characteristics make this segment of the economy/community all the more challenging, especially in light of new technology, trends in both centralization and decentralization, and new pricing strategies. Many advances are continuing at the Earthquake Engineering Research Centers, but more fresh approaches are needed (e.g., Grubesic et al., 2008).

• Development of planning frameworks for a cohesive set of policies. These would integrate structural and nonstructural initiatives in the urban landscape to avoid duplication, establish consistency, and capitalize on synergies.

• Exploration of equity and justice considerations. It is well known that earthquakes and other disasters have disproportionate effects across strata of society. The poor, minorities, the aged, and the infirm are more vulnerable, and even the middle class and those well off can be rendered indigent as a result of a disaster. Further study of the distribution of impacts of disasters is needed. Moreover an analysis of deeper equity and justice considerations are as well. For example, there is no consensus on the best definition of equity in any field, be it philosophy, political science, or economics (Kverndokk and Rose, 2008). An exploration of any unique aspects of the problem pertaining to earthquakes and other disasters is yet to be identified although some progress has been made (e.g., Schweitzer, 2006). Research would help call attention to this typically neglected area. Of course, it will take not only additional research but also political will to address it properly.

• Economic valuation methods have been used to measure cultural and historic "non-market" values in general (e.g., Navrud and Ready, 2002; Whitehead and Rose, 2009). Findings indicate individual willingness to pay to be rather low, but studies are needed to validate the methods and to identify the broader population for whom historic values are relevant, rather than just the property owners (Whitehead et al., 2008).

• Extension to ecological considerations. Some initial efforts appear to be promising (Renschler et al., 2007) in what looks to be an expanding field in terms of importance and research challenges. Additional work is also needed to explore the relationship between resilience (usually considered a short-run response) and adaptation to climate change (the long-run counterpart).

• All-hazards approach. Although this is typically acknowledged to be a valuable area of pursuit, research is still lagging since the last major advance (Mileti, 1999). Much of the work on terrorism, for example, has not yet been examined for application to earthquake resilience (e.g., NRC, 2006a; Vugrin et al., 2009).

• Long-run effects of earthquakes. We still do not have a definitive assessment of these implications. Part of the reason is the influence of external factors beyond the control of those who make the key decisions on reconstruction (e.g., business cycles, technological change, and globalization trends). It is also difficult to sort out the influence of external assistance vs. indigenous resources. Conceptual frameworks for more formal analysis of the issue should be encouraged (e.g., Chang, 2009, 2010).

• Resilience metrics indicators. It is important to be able to measure resilience in terms of both potential and practice. Basic metrics have been developed (Rose, 2004, 2007; Chang and Shinozuka, 2004) and applied successfully (e.g., Rose et al., 2009; SPUR, 2009), but more research is needed to include dynamic elements. More recent research on indicators of the various capacities needed to promote resilience (i.e., prospective resilience indicators) also appears promising (CARRI, 2010; Cutter et al., 2010). The relationship of performance-based engineering to resilience metrics is still unexplored.

- The evaluation and prediction of demand surge. Construction prices are likely to rise following a major earthquake. Although this is often attributed to the fact that there is an increased demand for repair and reconstruction, it also stems from the fact that construction equipment has been damaged, as have inventories of construction materials. Moreover, the production of even more materials may be limited because of damage to their manufacturers. This condition can raise the cost of recovery significantly. It involves an important tradeoff between recovering quickly at a high price and minimizing business interruption losses vs. incurring business interruption losses and waiting until prices settle down in order to reduce recovered costs. Theoretical and empirical analyses are needed to better understand this phenomenon and to be able to predict its course.

- Strategic planning. Now that much more emphasis has been placed on the post-disaster period, it is prudent to examine risk management across the entire timeline of earthquake disasters. This includes a closer examination of the relative payoff of resources expended before and after an event. In terms of reducing business interruption, post-disaster resilience appears to have an edge in terms of cost given the large number of low-cost alternatives and the fact that they need only be implemented once the event has taken place (unlike mitigation). Of course, mitigation is still the most relevant strategy for the prevention of building damage and the promotion of life safety.

- The post-disaster environment is an extreme one, where time appears compressed because of the pressures to restore normalcy (Johnson, 2009; Olshansky and Chang, 2009). The urban setting that might have taken decades or more to construct must be repaired or rebuilt in a much shorter time. However, much of the data that has been collected in past earthquakes will often have been effectively lost. If data collection is to become even more comprehensive, improved data management, archiving, and linking to existing data is required.

- Although leadership and management of post-disaster response-related activities are fairly well defined in the United States, the government's role in disaster recovery is less clear (NRC, 2006a). The disaster recovery process happens with the many simultaneous decisions made and resulting actions taken, by individuals, businesses, and institutions, both directly and indirectly impacted (Johnson, 2009). In turn, managing recovery should be about planning, organizing, and leading a comprehensive recovery vision, and influencing the many simultaneous decisions to achieve that vision as effectively and efficiently as possible (Johnson, 2009). Without a comprehensive understanding of the post-disaster needs or a recovery vision, bureaucratic management approaches often tend to be reactive, inflexible, and inefficient.

• Research has consistently shown that the post-disaster environment provides one of the most opportune times for mitigation and betterment, such as improved efficiency, equity, or amenity. The key influences that government can have are vision—often in the form of leadership and plans—and resources, most importantly money (Rubin, 1985; Johnson, 2009). But, recovery needs time to be accomplished thoughtfully and to allow for proper deliberation and public discourse on how to achieve risk reduction and betterments; and, recovery managers are often pressured to go faster than information, knowledge, and planning generally flows. Furthermore, the amount and flow of money often doesn't match the compressed pace of recovery or work efficiently and effectively to achieve substantial, long-term risk reduction. It also is not flexible enough to address the evolving needs in a post-disaster environment.

• In recent years, researchers have proposed some guiding principles to manage the complexities of post-disaster response and recovery and help reduce disaster-related costs and repetitive losses. Post-disaster resilience—as one of these guiding principles—means that there is more decentralized and adaptive capacity in affected communities to effectively contain the effects of disasters when they occur and manage the recovery process, as well as an ability to minimize social disruption and mitigate the effects of future disasters (Sternberg and Tierney, 1998; Bruneau et al., 2003; NRC, 2006a).

Enabling Requirements—Methods and Models

Improved data and methods are needed to analyze and promote resilience, both pre- and post-disaster. This includes more extensive data collection, building on the work by Cutter and Mileti (2006). The data should be made available through established clearinghouses, such as the Natural Hazards Research Applications and Information Center (NHRAIC) at the University of Colorado at Boulder and the Community and Regional Resilience Institute (CARRI), and new knowledge hubs such as the proposed National Center for Social Science Research on Earthquakes and Other Disasters (Task 8 above). It is essential that the PIMS in Task 9, above, include data on social and economic consequences.

Many sound economic models exist to measure the losses from natural hazards before and after the implementation of resilient actions. Input-output (I-O) analysis, with a few exceptions (e.g., Gordon et al., 2009), is not up to the task because of its inflexibility, (i.e., inability to readily incorporate aspects of substitution between inputs, domestic productions and imports, changes in behavior, etc.).[23] Mathematical programming (MP) models overcome some of the limitations of I-O and are useful where spatial analysis or technical details are important (e.g., Rose et al., 1997).

Computable general equilibrium (CGE) analysis captures all of the advantages of I-O (e.g., multi-sector detail, total accounting of all inputs, focus on interdependence) but overcomes its limitations by including behavioral considerations, allowing for substitution and other nonlinearities, and reflecting the workings of markets and prices (Rose, 2005; Rose et al., 2009). Macro-econometric models are coming into increasing use in the analysis of the impacts of the economic consequences of earthquakes and other disasters (Rose et al., 2008). They have the advantage over CGE models in being based on time series data, rather than mere calibration of a single year of data, and provide a basis for forecasting the future. They are also superior thus far in incorporating financial variables, although this is not necessarily an inherent advantage of the modeling approach.

Agent-based models are increasingly in vogue in disaster studies. They examine individual motivations and their actions within groups. They are especially adept at simulating panic, contagion, and aversion behavior (e.g., Epstein, 2008). They have more recently been expanded to analyze aspects of urban form, and would therefore be useful in analyzing people's location decisions under alternative zoning and broader land-use restrictions (e.g., Heikkla and Wang, 2009). Finally, systems dynamic models represent an excellent overall framework for the combination of various types of models into an overall framework. It is unlikely that any single modeling approach is best for all aspects of an earthquake or other disaster; the trick is often being able to integrate several different models successfully (e.g., Chang and Shinozuka, 2004). In effect, several integrated assessment models of earthquake risk, vulnerability, and consequences (e.g., HAZUS) are a subset of systems dynamic models.

Many potential modeling candidates continue to flourish, but the research community and practitioners have been rather slow to validate them (e.g., Rose, 2002). Further work on validation would inspire confidence in the use of many valuable models.

Supplementing overall model validation, individual hypotheses contained in many of these models relating to individual and group behavior need to be tested. In addition to those noted elsewhere in this section, especially urgent are hypotheses about changes in individual risk perceptions due to the social amplification of risk. They have been shown in prelimi-

---

[23] The measurement of resilience is contextual, i.e., it requires a reference point of what the world would look like without resilience. Ironically, I-O analysis is ideal for this purpose in that it best mimics an inflexible rather than a resilient response.

nary analyses to vary by type of threat and other factors (Burns and Slovic, 2009) but in general can lead to economic behavior that greatly exacerbates business interruption losses (Giesecke et al., 2010). Also key will be new studies on the public reaction to improved earthquake prediction.

Although HAZUS (FEMA, 2008) represents a major milestone in making it possible for many analysts to undertake hazard loss estimation, it is not without its limitations. Only a very small proportion of its funding has been devoted to the Direct and Indirect Economic Loss Modules (DELM and IELM). Hence, it is not surprising that HAZUS contains major gaps and shortcomings.

One of the most glaring is the inability of HAZUS to estimate the majority of economic losses from utility lifeline disruptions. HAZUS provides only an estimate of the damage to utility system components. No capability is provided in the DELM to evaluate the much larger disruption to the first round of utility customers. This problem is magnified, because there is input into the IELM to evaluate the further ripple effects throughout the economy.

We acknowledge that this is a challenging subject largely because of the complex network characteristics of electricity, gas, water, transportation, and communication lifelines. However, we encourage the study of the feasibility of incorporating network data and computational algorithms into HAZUS. We also suggest the consideration of a HAZUS "patch" that can be added to the software as an interim remedy. Such a construct has been developed as part of the report to Congress on the benefits of FEMA hazard mitigation grants (MMC, 2005; Rose et al., 2007), and further refined in a contribution to the Risk Analysis and Management for Critical Asset Protection (RAMCAP) loss estimation software (Rose and Wei, 2007). These patches make use of the input-output data on the lifeline requirements of the sectors of the economy and provide a reasonable approximation of the direct and indirect economic impacts and without a full network capability.

Other aspects of HAZUS that could stand refinement pertain directly to resilience:

- The recapture factors in the DELM are labeled as being applicable for periods up to 3 months, but no guidance is provided for longer time periods. The possibility of recapture of lost production wanes over time as customers seek other suppliers.
- Potential business relocation is implicit in the DELM estimates, but is mentioned only in passing in the Users Guide. The user has no way of ascertaining the proportion of impacts that can be avoided through average relocation practices.
- The IELM contains several sources of resilience, including import

substitution and excess capacity. Implementing these options can result in extreme play in the model, i.e., small changes in parameter values can yield extreme changes in impacts. More extensive analysis is needed to determine the extent to which the ensuing results are overly sensitive to the parameter's changes and limitations of the user's abilities.

Although HAZUS serves as a useful tool for loss estimation, it is necessarily simplified to render it accessible to a broad range of users, as it is intended to be accessible at every emergency planning office in the United States. Myriad advances have been made in more sophisticated, and typically more accurate integrated assessment models, especially at the various NSF Earthquake Engineering Research Centers (e.g., Shinozuka et al., 1998). These represent major advances in the state of the art conceptually and empirically that provide greater insights into the loss and recovery process. The pace of this research has recently declined because of the end of NSF support of these centers following their "graduation." Private industry support has always been lacking because of the double "public good" nature of this research. First, it focuses a great deal on infrastructure, which provides benefits to all of society, and from which it is difficult to extract a payment for all of its services. Moreover, the gains from any single research effort exceed the value to any one sponsor.

To provide enhanced leadership for reducing long-term risk and repetitive losses, post-disaster, will require a commitment and leadership among the NEHRP agencies to work with other federal agencies that have responsibility for many of the federal programs that provide funding to communities to recover from an earthquake or other damaging disaster.

Implementation Issues

The implementation issues presented here overlap in part and are generally consistent with Tasks 8 and 11.

NEHRP agencies must help ensure that post-disaster federal programs contain support to increase the earthquake resiliency of the impacted communities. For example, FEMA regulations should promote increased application of mitigation funding under the 406 Public Assistance Program.

The lead agency and staff for NEHRP must also ensure that the intents of this action are integrated with the recommended community-level capacity-building initiative also proposed as part of this study. Pilot city projects (see Task 18) could be initiated post-disaster in communities affected by damaging earthquakes or other disasters. The post-disaster environment would provide an excellent opportunity to develop and test mitigation tools at the community level. Efforts must also be made to provide both pre- and post-disaster mitigation planning for recovery.

Having communities prepared to make recovery decisions and take action can result in a better organized and executed local recovery. Savings from such efforts are likely to rival savings that can be achieved by structural measures alone.

It is not a requirement that the federal government design, fund, direct, and implement all resilience activities; rather, stronger working relationships and enlightened multi-level governance will further promote resilience in practice.

TASK 11: OBSERVATORY NETWORK ON COMMUNITY RESILIENCE AND VULNERABILITY

The research community has, on numerous occasions, identified key data collection obstacles that are impeding the advancement of knowledge regarding earthquakes and other disasters, their impacts on society, and factors influencing communities' disaster risk and resilience (see also Task 8) (Mileti, 1999; NRC, 2006a; Peacock et al., 2008). In particular:

1.   Current funding mechanisms do not allow monitoring of long-term changes in disaster resilience and vulnerability. Disaster research has traditionally been supported by "quick-response" grants to gather ephemeral post-disaster data, standard 3-year research grants, and research centers that are typically funded for 5 years. Although these mechanisms have led to significant advances in knowledge, they preclude longer-term monitoring over multi-year or decadal timescales.

2.   Standardized data collection protocols do not exist. Data collection efforts by individual investigators rarely replicate measurement tools and methods applied in previous studies. Consequently, although substantial amounts of data have been collected on numerous disaster events, the ability to compare across events and studies, replicate findings, and draw generalized conclusions is very limited.

3.   Effective mechanisms for coordinating and sharing data among researchers and practitioners are largely lacking. Issues relate not only to long-term data repositories but also to intellectual property, confidentiality, data archiving protocols, etc. Data that are currently collected are not shared and utilized as widely as they could be.

4.   A comprehensive, holistic view of community disaster vulnerability and resilience is beyond the purview of individual research studies. Individual studies, for reasons of practicality, largely focus on specific, limited aspects of risk; however, understanding and fostering community resilience requires comprehensive knowledge that builds on and synthesizes across many studies. For example, understanding how community

resilience varies with the size of the earthquake event is an important knowledge gap that requires synthesis across many studies.

Investment in the research infrastructure to overcome these barriers would be strategic and highly cost-effective. It would catalyze rapid advances in the knowledge needed for disaster resilience—facilitating the development of systematic and cumulative rather than piecemeal knowledge bases, addressing fundamental knowledge gaps, and enabling development of models that explain changes in community vulnerability and resilience over time.

Proposed Actions

In the next 5 years, NEHRP should establish a network of observation centers—an "Observatory Network"—to *measure, monitor, and model* the disaster vulnerability and resilience of communities across the nation. This observatory network would focus on the dynamics of social systems and the natural and built environments with which they are linked. The network would facilitate efficient collection, sharing, and use of data on disaster events and disaster-prone communities. Its activities—including standardization of data collection protocols, archiving of data, long-term monitoring of community vulnerability and resilience (especially in high-risk regions), and regular reporting of these assessments—would provide critical research infrastructure for advancing and implementing knowledge to reduce disaster risk nationwide.

An Observatory Network would focus on data collection related to four principal thematic areas: (1) resilience and vulnerability; (2) risk assessment, perception, and management strategies; (3) mitigation activities; and (4) reconstruction and recovery.

Existing Knowledge and Current Capabilities

The state of knowledge in disaster social sciences with respect to the thematic areas that would be addressed by an Observatory Network has been thoroughly documented in several reports. For comprehensive overviews of research progress, see in particular: the five-volume second assessment of research on hazards and disasters (Mileti, 1999; together with companion volumes, Burby, 1998; Kunreuther and Roth, 1998; Cutter, 2001; and Tierney et al., 2001), and a report by the National Research Council that documented social science contributions under NEHRP (NRC, 2006a). (See also discussions related to Tasks 8 and 10 in this report.) Here, a few examples of knowledge advances and important remaining gaps are provided to illustrate the potential benefits of an Observatory Network on community vulnerability and resilience.

Vulnerability and resilience represent core concepts in understanding the effects of hazard events on populations. Vulnerability broadly encompasses exposure (e.g., number of people living in seismic regions) and physical vulnerability (e.g., building design and construction), as well as social vulnerability (e.g., access to financial resources to cope with an earthquake). Resilience, as noted in Chapter 2, refers to the ability to reduce risk, maintain function in disaster events, and recovery effectively from them. Key research questions include: Which communities are most vulnerable to disaster losses, and why? How can we measure and assess disaster resilience? How are vulnerability and resilience changing over time? What accounts for these patterns and changes?

A substantial knowledge base has developed regarding disaster vulnerability and, increasingly, disaster resilience. Empirical studies have, however, been dominated by one-time case studies limited in time and space. One exception is work by Cutter et al. (2003) that systematically assessed an index of social vulnerability, using 42 census variables to identify clusters of highly vulnerable counties across the United States. It is now necessary to extend this type of work to more holistically assess vulnerability and resilience; in particular, by integrating social vulnerability with patterns of exposure, physical vulnerability, and hazard probabilities, by considering interactions with ecological health, infrastructure, and institutional and community capacity, and by tracking changes over time (Cutter et al., 2008b). This will require access to local data on such important factors as local building codes, their enforcement practices, and seismic upgrading or other mitigation programs. In contrast to census data, such information will be highly variable from one jurisdiction to the next. Moreover, such information will likely be gathered by many research teams in the context of diverse, locally oriented projects. Thus it is imperative for the research community to develop and utilize standardized, common data collection protocols and instruments, and to share the data acquired in a shared repository.

Knowledge regarding *risk assessment, risk perception, and management strategies* is also fundamental to advancing national earthquake resilience. Risk assessment refers to estimates by technical experts on the likely consequences of potential hazard events, typically deriving from formal probabilistic models. (See also Tasks 6 and 7 regarding earthquake scenarios and earthquake risk assessment.) How risk is perceived by individuals, organizations, and groups within society, however, often differs from experts' assessments of risk. As noted by Peacock et al. (2008, p. 9), "risk management strategies . . . require the development of policies that take into account both risk assessment and perception and include economic incentives (e.g., subsidies and fines), insurance, compensation, regulations (e.g., land use restrictions) and well-enforced standards (e.g., building codes) . . . and . . . will often require private-public sector partnerships." Central research questions include: How do the risk perceptions of stakeholders (e.g., residents, community leaders, elected officials) differ across communities and over time? How does the actual experience of disasters affect risk perception and change risk-related behaviors? How long do these changes last? How prevalent is insurance among different communities and groups within communities, and how does this change over time? What explains differences in pro-active behaviors?

Substantial progress has been in social scientific understanding of such questions. A few attempts have been made to develop insights from multiple, rather than single, case studies; for example, Webb et al. (2000) conducted consistently designed, large-scale surveys of businesses in five communities, one of which had not experienced a major disaster (Memphis/Shelby County, TN) and four of which had (Des Moines, IA; Los Angeles, CA; Santa Cruz County, CA; and South Dade County, FL). Their inquiry included business disaster preparedness, as well as loss and recovery. Although this series of studies was ground-breaking, it was nonetheless cross-sectional rather than longitudinal in design, "making it impossible to track changes that occurred over time" (Webb et al., 2000; p. 89). Findings were also highly variable across cases, suggesting the need for many more cases to better understand patterns of difference. The study focused on organizational, agent-specific, and community factors that contribute to disaster preparedness (e.g., business size and disaster experience). Results suggested that these factors alone provide only a partial explanation of behavior; other multiple and complex influences, including for example the decision-making processes of business owners, must also be investigated. These limitations—lack of longitudinal perspectives, small samples of communities, and partial rather than holistic explanations of risk-related behaviors—remain important knowledge gaps.

Pre-disaster mitigation activities—including addressing seismic risk through building codes, structural and nonstructural retrofits, and land-use planning—represent the primary means through which earthquake losses can be reduced in the long term. Fundamental research questions include: What factors influence the adoption of mitigation measures by households, businesses, and communities? To what extent have mitigation activities been adopted in different communities? What types of mitigation measures are most cost-effective? How can insurance and other programs promote mitigation? How effective are plans and planning processes (e.g., the state mitigation plans that are required, and the local mitigation plans that are encouraged, under the Disaster Mitigation Act of

2000) in reducing vulnerability and increasing resilience? How do different types of legal and legislative contexts influence mitigation activities?

Although a number of studies have addressed mitigation decision-making at various scales, from individuals to state governments, relatively few have rigorously assessed the effectiveness and cost-effectiveness of mitigation. In one key study, the Multihazard Mitigation Council recently conducted a congressionally mandated, independent study to assess the future savings from various types of mitigation activities (MMC, 2005; see also Task 10 above). Focusing on data for 1993 to 2003, this landmark study found that FEMA's natural hazard mitigation grant programs were cost-effective and resulted in considerable net benefits in the form of reduced future losses from natural disasters. On average, every $1 of FEMA mitigation grant funding led to societal savings of $4. Yet further knowledge about the cost-effectiveness of specific mitigation approaches is still needed for informing policy. The report concluded that (MMC, 2005, p. 6):

> Continuing analysis of the effectiveness of mitigation activities is essential for building resilient communities. The study experience highlighted the need for more systematic data collection and assessment of various mitigation approaches to ensure that hard-won lessons are incorporated into disaster public policy. In this context, post-disaster field observations are important, and statistically based, post-disaster data-collection is needed for use in validating mitigation measures that are either costly, numerous, or of uncertain efficacy or that may produce high benefit-cost ratios.

Reconstruction and recovery represent a fundamental dimension of disaster resilience, yet it is widely acknowledged to be the least-understood phase of the disaster cycle (Tierney et al., 2001; NRC, 2006a; Peacock et al., 2008; Olshansky and Chang, 2009). Key questions concern: Why do some communities recover more quickly and successfully than others? How does the recovery trajectory of communities differ by type and magnitude of the hazard event, conditions of initial damage, characteristics of the community, and decisions made over the course of reconstruction and recovery? Who wins, and who loses, in the process of disaster recovery? How do different types of assistance and recovery resources affect household and business recovery? What types of decisions and strategies are most critical to recovery? How do disasters affect communities over the long term?

Current knowledge about post-disaster recovery has been developed through case studies of individual disasters, and systematic data collection is greatly needed: "Indeed, without sufficient data on short and long term recovery with respect to households, housing, businesses, and other components of our communities, developing and validating models of community resiliency or assessing the effectiveness of recovery policy and planning will remain elaborate conjectures" (Peacock et al., 2008, p. 11). One study has attempted to develop a prototype computer model of community disaster recovery (Miles and Chang, 2006; see also Olshansky and Chang, 2009), accounting for many of the factors and interactions suggested in the case study literature. The model simulates how recovery trajectories might vary in different conditions. The study found, "The paucity of data and empirical benchmarks is a major challenge. There are not enough disaster events that have been systematically studied from the perspective of developing quantitative data and recovery indicators. . . . Moreover, data are hard to come by, often inconsistent and incomplete, and typically expensive to gather" (Olshansky and Chang, 2009, p. 206).

Enabling Requirements

An Observatory Network on community resilience and vulnerability is needed, analogous to observatory networks that have been established in the environmental science arena. A prime example is the Long-Term Ecological Research (LTER) network established by NSF in 1980. The LTER network currently includes 26 sites located in a diversity of ecosystems, from Antarctica to the Florida Everglades.[24] Across these landscapes, researchers have been investigating similar scientific questions, sharing data, and synthesizing ecological concepts. In recent years, NSF has also established the National Ecological Observatory Network[25] (NEON) to provide infrastructure, as well as consistent methodologies, to support research on continent-scale ecology related to climate change impacts and other large-scale changes. It is also supporting the Collaborative Large-Scale Engineering Analysis Network for Environmental Research (CLEANER) (NRC, 2006c). Within the earthquake field, NEES represents a large-scale investment in a nationally distributed network of shared engineering facilities for experimental and computational research.

A national observatory network is needed to address the disaster vulnerability and resilience of human communities (e.g., cities), using methodologies applied consistently over time and space, with attention to the complex, place-based interactions between changes in social systems, the built environment, and the natural environment. In June 2008, USGS and NSF sponsored a workshop to outline the goals, research agenda, data collection principles, structure, and implementation of such a network, labeled the Resiliency and Vulnerability Observatory Network (RAVON) (Peacock et al., 2008). Participants at that workshop agreed that the network should focus on natural disasters, foster interdisciplinary

---

[24] See www.lternet.edu.
[25] See www.neoninc.org.

research, facilitate comparative research, and emphasize issues of social vulnerability. The committee supports the establishment of such a network and considers it to be a high priority for implementing the 2008 NEHRP Strategic Plan in the next 5 years.

An Observatory Network would consist of a series of research nodes, comprising at least three types:

• *Regional nodes.* Regional hubs for coordinating researchers (from both within and outside the region) who are gathering data about a specific geographic region. Such a hub could coordinate activities and work closely with local governments, nongovernmental organizations, and community groups in the region. These nodes could be strategically located in disaster-prone regions of the country, pre-positioned to facilitate rapid post-disaster studies of communities struck by natural disasters.

• *Thematic nodes.* Existing centers whose missions directly relate to the Observatory Network could be included as thematic nodes. These could include research centers (e.g., the Natural Hazards Center at the University of Colorado at Boulder, which already provides information clearinghouse services and convenes annual workshops for the research and practice communities) as well as mission agencies such as USGS that already develop spatial databases on hazards and risks nationwide. The proposed National Center for Social Science Research on Earthquakes and Other Disasters (Task 8 above) would also serve as a key node.

• *Living laboratory nodes.* Nodes could be established in communities affected by major natural disasters, in order to gather and assess data on disaster impacts and recovery over the long term.

Core activities of such a network would include:

• Developing and sharing standardized definitions, measurement protocols, instruments, and strategies for data collection across multiple communities and disasters;

• Developing and archiving longitudinal databases for analyzing and modeling resilience and vulnerability over time;

• Supporting researchers investigating new disaster events.

These activities would be closely linked with post-earthquake social science research on response and recovery (Task 8), data collection and sharing capabilities of the Post-Earthquake Information Management System (PIMS) (Task 9), and socioeconomic research on mitigation and recovery (Task 10).

The scope of the Observatory Network should be multi-hazard, including but not be exclusively focused on earthquake hazard. It is assumed that many of the nodes of the Observatory Network will be geographically distributed across regions of the United States with significant earthquake hazard. Some nodes, however, may be located in regions of low seismic risk but high risk of hurricanes, floods, or other hazards; for example, a "living laboratory" node could be established in a region recovering from a catastrophic hurricane.

This multi-hazard emphasis is advantageous for two reasons. First, because major earthquakes occur infrequently, it is important to take advantage of lessons from other types of disasters. In contrast to some domains of technical knowledge (e.g., seismology or earthquake engineering), for issues related to communities' social and economic vulnerability and resilience, many commonalities exist between earthquakes and other hazards. Knowledge about earthquake vulnerability and resilience could thus be advanced more rapidly through multi-hazard research than through an exclusive focus on earthquakes. At the same time, data and knowledge on earthquake vulnerability and resilience should be utilized to advance communities' resilience to the other types of hazards that many earthquake-prone communities also face. The Observatory Network should therefore be structured to foster data sharing, comparative study, and policy analysis across hazards.

Implementation Issues

Implementation of a distributed network is likely to require a multi-year, phased process. This implementation may involve a number of challenges and associated opportunities:

- Developing an effective governance and decision-making structure;
- Developing an effective phased implementation;
- Integrating with existing research centers and organizations engaged in data collection on disaster vulnerability and resilience, and avoiding duplication of effort;
- Locating and building new nodes of the network;
- Developing widely accepted, standardized data collection instruments and protocols;
- Addressing tensions between individual investigator-led research and the need for commonalities across studies;
- Developing effective infrastructure and protocols for data archiving and sharing, including addressing issues of data confidentiality and intellectual property;
- Developing a sustainable mechanism for long-term financing of the network;
- Bridging research findings with disaster risk reduction practice.

The committee supports the overall implementation structure suggested in the USGS- and NSF-supported RAVON workshop (Peacock et al., 2008). That workshop suggested that in the first 5 years:

- Phase I: an initial 5-6 nodes would be established (at $400,000 per year) through a competitive process;
- Phase II: a steering committee is established, comprising the principal investigators of the Phase I nodes. The steering committee would develop the network's charter and constitution, and assist NSF in selecting additional nodes. Technical subcommittees would develop data collection and associated protocols. Approximately 3 to 5 additional nodes, together with a network coordinating grant, would be established after open competition.

After the first 5 years, the network steering committee would have the option of recommending a Phase III for additional network growth.

TASK 12: PHYSICS-BASED SIMULATIONS OF EARTHQUAKE DAMAGE AND LOSS

The goal of physics-based simulations of earthquake damage is to replace uncoupled, empirical computations of earthquake shaking, nonlinear site and facility/lifeline response and damage, and loss, with fully coupled numerical simulations (so-called end-to-end simulations) that use validated numerical models of materials and components. The purpose is to greatly improve the accuracy of, and reduce the uncertainty in, earthquake response, damage, and loss calculations of new and archaic elements and systems in our built environment.

Proposed Actions

➤ Advance the practice of engineering design practice across all disciplines through the development and implementation of validated multi-scale models of materials, components, and elements of the built environment, and the use of high-performance computing and data visualization.

➤ Maximize the impact on national earthquake resilience by *integrating* knowledge gained in Tasks 1, 13, 14, and 16 and *"operationalizing"* the integrated product using end-to-end simulation.

Existing Knowledge and Current Capabilities

Current and planned tools for performance-based earthquake engineering of the built environment involve a series of uncoupled analyses of earthquake ground motion, site (soil) response, and foundation and structural response as follows:

• Empirical predictive ground motion models are used to estimate the effects of earthquake shaking (measured using a spectrum) at a rock outcrop below the facility or lifeline. These models aggregate the effects of P, SH, and SV waves; 6-component acceleration times series are not developed.

• Site response computations typically involve the vertical propagation of SH waves (ground shaking) from a rock outcrop to the free field, including the ground surface. The SH wave time series input to the soil column are matched to the rock outcrop spectrum. The layers of soil in the column are modeled using equivalent linear properties. The output of the site response analysis is a response spectrum in the free field.

• Engineering calculations are performed using *empirical* models of materials (e.g., soils, metals, concretes, polymers) and simplified macro models of components (e.g., rolled steel beams and reinforced concrete columns). Models of materials and components have been developed on the basis of regression analysis of a limited body of test data with limited understanding of the underlying physical processes. Unified models for materials subjected to arbitrary mechanical and thermal loadings do not exist (but are proposed for development in other tasks).

• Facility and lifeline response (damage) computations are typically performed using empirical macro models of structural components assembled into a numerical model of the facility/lifeline and spectrum-compatible earthquake ground motions input to the numerical model at the ground surface. Nonstructural components and assemblies are treated as cascading systems. Fragility functions for structural and nonstructural components use simplified demand parameters of maximum (lateral) story drift and peak horizontal floor acceleration.

• Losses are computed at the component level using maximum computed responses, fragility data, and consequence functions, and simply aggregated across the breadth and height of the facility/lifeline.

• Seismologists and earthquake engineers have done extensive research on simulating fault rupture and seismic wave propagation through rock, soils, and built structures, and these efforts are beginning to be coupled into end-to-end simulations (Muto et al., 2008). However, *empirical* linear models of soils and structures/facilities/lifelines have generally been used in these calculations, and better models that incorporate nonlinear effects are needed. Computational frameworks to accommodate physics-based simulations are being developed by SCEC and other organizations; such cyberinfrastructure is an enabling technology for this task.

Enabling Requirements

Robust multi-scale nonlinear models are the future of engineering science. The coupling of these models from the point of rupture through rock and soil into structure represent the future of professional design practice. The implementation of fully coupled, nonlinear macro models of geological and structural components and facility framing systems will enable the profession to move beyond empirical models of unknown reliability. The basic science and engineering knowledge needs required to develop and implement these next generation models, tools, and procedures include the following items:

• Earthquake generation and propagation through heterogeneous underground structure is a very complex process that is impossible to observe. A comprehensive research program is needed to better characterize earthquake sources and the strong ground motions they produce, as described under Task 1.

• Facilities or lifelines are often constructed on a 3D heterogeneous soil basin of varying depth and breadth. Each basin experiences body waves propagated from the rock on the boundaries of the soil basin, where the waves on the boundary vary as a function of location and time. Each basin develops surface waves as the body waves strike the earth's surface. The soils in the basin are highly nonlinear and may flow (liquefy) dependent on the amplitude and duration of the shaking. As described in other tasks, NSF and USGS should develop techniques to map heterogeneous geologic structure at depth and multi-scale and multi-phase models of soils and rock to enable the propagation of seismic wave fields from the earthquake source to the facility/lifeline. Multi-scale and multi-phase models should be validated from large-scale tests using NSF-NEES infrastructure.

• Facility or lifeline response (damage) is dependent on many factors including depth of embedment in the soil, the spatial and temporal distribution of the body and surface waves across the plan dimensions of the facility/lifeline, seismic wave scattering, the size and mass of adjacent structures, the numerical models of structural and nonstructural components used to compute response, and the damage functions assigned to each component. As described in other tasks, NSF should fund the development of multi-scale constitutive models for archaic, modern, and new high-performance materials, and translate these constitutive models into validated hysteretic macro models of components using advanced numerical tools and data from tests of large-size components using NSF-NEES infrastructure, as described in Tasks 13 and 14.

• The calculation of casualties, repair cost, and business interruption requires complete information on the distributions of damage to structural and nonstructural components through failure, and the consequences (casualties, repair cost, and business interruption) of the damage and how the consequences aggregate across the breadth and height of the facility/ lifeline. FEMA should fund the development of component-level fragility functions for archaic and modern components of buildings, bridges, and infrastructure by a combination of numerical studies and large-scale testing using NSF-NEES infrastructure. Robust strategies to assemble component-level consequences into estimates of system-level loss must be prepared.

• Fully coupled physics-based simulations must include a formal treatment of uncertainty and randomness. Complete Monte Carlo simulations will be prohibitively expensive in terms of computational effort. Efficient numerical techniques must be developed for end-to-end simulations.

• Any fully coupled, physics-based simulations from the point of rupture to the response of structural and nonstructural components in a facility of lifeline will be computationally expensive. NSF should continue to develop high-performance computing capabilities in support of such simulations for future use by researchers, design professionals, and decision-makers.

• Fully coupled, physics-based simulations will generate terabytes, and even petabytes, of data. NSF and USGS should develop analysis and visualization tools to process large volumes of data and enable decision-making in a timely manner.

Implementation Issues

Physics-based simulations describing the response of the built environment to fault rupture have been deployed on a limited basis (Olsen et al., 2009; Graves et al., 2010). The constitutive models used for soil and structural components are empirical and linear. The replacement of empirical linear models with multi-scale nonlinear models as described in other tasks will represent a paradigm shift in engineering science and practice in the United States. Successful execution of this task is contingent on funding of the basic science and engineering described in Tasks 1, 13, 14 and 16.

A major challenge facing the implementation of fully coupled, physics-based simulations is the interdisciplinary education of the next generation of engineers and scientists, who will have to be expert in earth science and physics, engineering mechanics, geotechnical engineering, and structural engineering, to be qualified to perform these simulations. University curricula will have to be changed. Research results will have to be disseminated quickly to academics for inclusion in graduate-level coursework.

TASK 13: TECHNIQUES FOR EVALUATION AND RETROFIT OF EXISTING BUILDINGS

Buildings built without adequate consideration of the earthquake effects that are appropriate for their location dominate the nation's exposure to earthquakes. These buildings may be seismically vulnerable because they were built before seismic codes were enforced in their region, because the codes used were not yet mature, or because the earthquake threat is now known to be greater than when they were designed. Further the design provisions used for buildings that are critical for post-earthquake response and recovery (e.g., hospitals, fire stations, emergency operations centers) may not provide adequate performance for their intended function. Although lifelines are critical for community resilience, the cost of damage of buildings and their contents and the resulting business interruption or downtime typically account for the bulk of the total economic loss from a large-magnitude earthquake. Further, the greatest threat to life loss in earthquakes in the United States is posed by existing buildings. However, the replacement or retrofit of a significant portion of our vulnerable building stock is not practical in the short or medium term. The current assessment methodologies cannot identify buildings whose performance may prevent the desirable level of resilience for individuals or communities with sufficient accuracy to develop efficient overall mitigation programs. Similarly, current retrofit design standards, although performance-oriented, coupled with currently employed construction techniques, may not provide the targeted performance with adequate reliability and economy.

It is critical in the near term that assessment methods are refined to more accurately identify buildings with inadequate performance and to provide engineers with the tools and procedures to economically retrofit those so identified. The development of Next-Generation Performance-Based Design in the ATC-58 project[26] is expected to provide such capabilities, but it will be 10 years or more before this methodology is refined sufficiently to provide standards for evaluation and retrofit for widespread use. Many of the activities in this task will also apply to Task 14.

Proposed Actions

➢ Conduct integrated laboratory research and numerical simulations to substantially increase understanding of the nonlinear response of archaic materials, structural components, and framing systems.

➢ Develop reliable, practical analytical methods that predict the response of existing buildings with known levels of reliability.

---

[26] See www.atcouncil.org/.

➤ Improve consensus standards for seismic evaluation and rehabilitation to improve effectiveness and reliability, particularly with respect to predicting building collapse.

➤ Develop simplified methods of evaluation of known reliability.

Existing Knowledge and Current Capabilities

The issue of seismic risk from existing buildings was put on a national stage when FEMA launched its Program to Reduce the Seismic Hazards of Existing Buildings. The Action Plan describing the main elements of that program was developed at a workshop held in Tempe, AZ, in 1985 (FEMA, 1985a, 1985b). Although this program generated many intermediate useful documents (reference list), its initial goals were reached with the publication of a national seismic evaluation standard, ASCE 31 (ASCE, 2003), and a standard for seismic rehabilitation, ASCE 41 (ASCE, 2007). Although these standards are now referenced in building codes and are widely used here and abroad, it is well recognized that sufficient knowledge of the nonlinear performance of many archaic materials, components, and framing systems is not available, and much of the component modeling and acceptance criteria in these standards is based on engineering judgment. Perhaps more concerning are results of several multiple-building evaluations using ASCE 31 that indicate 70 percent to 80 percent of older buildings fail the seismic life safety standard (Holmes, 2002; R&C, 2004). Limited earthquake experience in the United States in the past 60 years does not confirm these results and, importantly, it is not practical to replace or retrofit 70 percent to 80 percent of our older building stock.

In addition, there are indications that retrofits conforming to the ASCE 41 standard are overly expensive and may be conservative. Studies are currently under way at the National Institute of Standards and Technology to calibrate retrofits resulting from the provisions of ASCE 41, with respect to design requirements for new buildings. Other recent studies (FEMA, 2005) have shown that the use of the primary analysis method recommended in ASCE 41, depending on the "pushover" technique, should be limited to low-rise buildings, possibly not to exceed three stories. Although the next update cycle for ASCE 41 has begun, these major shortcomings cannot be rigorously addressed without developing a body of data from tests of large-scale components and systems, new numerical models of structural components that are suitable for computer analysis of existing buildings, and assessment procedures that are both efficient and sufficient for predicting the seismic response of archaic construction from incipient damage through collapse.

Rehabilitation techniques have been refined for some archaic structural framing systems and a few innovative techniques such as the use of fibre-reinforced plastic (FRP) wraps and coatings and the addition of damping have been developed, but cost and business disruption remain major deterrents to the widespread seismic rehabilitation that is needed to achieve a substantial reduction in our nation's earthquake risk. Retrofit costs are commonly of the same magnitude as the entire structure of a new building and for special cases such as historic buildings can be several times more than the structural cost of a new building. The most common retrofit techniques are described in a FEMA document, *Techniques for the Seismic Rehabilitation of Existing Buildings*, FEMA 547 (FEMA, 2006), but more complete training of engineers and other stakeholders about the entire retrofit process will be required to accomplish significant reduction of risk from existing buildings, particularly in regions with risks less clear than from the known faults on the West Coast.

Enabling Requirements

Many of the basic knowledge needs and implementation tools required to improve this significant mitigation activity are similar to those needed to advance performance-based earthquake engineering. However, recent studies by the community have identified activities specifically related to existing buildings. ATC 73 (ATC, 2007) contains recommendations for research, and ATC 71 (ATC, 2009a) presents an action plan for implementation. The following recommendations draw heavily from these two publications:

• Establish a coordinated research program related to existing buildings. The NEES facilities provide most of the hardware to accomplish the physical testing needed, but such a program requires resources much greater than those currently supported by NSF. Numerical simulations must be integrated with physical testing, requiring additional support from NSF, other federal agencies, city and state agencies, and industry.

• Develop fragility data for structural and nonstructural components of existing buildings, both to support the development and utility of performance-based earthquake engineering in the long term and to improve decision-making using current retrofit procedures in the short term.

• Improve collapse-prediction capabilities.

• Undertake full or large-scale shake table tests of complete building structural and/or nonstructural systems.

• Perform extensive in-situ testing of existing buildings and components thereof. Many opportunities to test buildings scheduled for demolition have been missed. A program targeted at such testing, with incentives to building owners, should be started.

- Current analytical procedures often predict higher probabilities of failure than have been observed in post-earthquake reconnaissance, particularly for short period buildings. Soil-foundation-structural interaction effects on input shaking intensity may help explain these observations. Addition study in this area, particularly lateral decoupling of the structure and ground, is needed.
- Develop retrofit methods that are more targeted at specific deficiencies and are less costly and intrusive than current standards of practice. There are on-going projects at the time of this writing that seek to identify deficient buildings and quantify their exposure, including EERI's Concrete Coalition, and research projects funded by NSF under the NEESR program. The results of these projects must be carefully mined to update prior estimates of the vulnerability of the national building stock.
- Develop techniques for nondestructive testing of in-situ structural materials and components and methods of creating the as-built concealed geometric data needed for analysis.
- Develop a comprehensive, nonproprietary building rating system, building on the ATC-50 project that developed such a system for wood frame buildings. Such a system has been recommended to bring seismic safety "into the marketplace" for decades and is discussed at virtually every workshop related to seismic safety. Analytical tools to create a reliable rating system may soon be available. Other issues related to administration of such a system and quality control must also be resolved.
- Develop a uniform method of translating test data to acceptance criteria for use with current analysis and retrofit design methods.
- Collect, curate, and archive building inventory data in all seismic regions to facilitate regional loss estimation and to focus research on the most common high-risk building and structural types.
- Calibrate evaluation methodologies and prediction of damage states both with earthquake damage data and with performance expectation of new buildings.
- Calibrate retrofit standards and techniques with performance expectations of new buildings.
- Clarify structural and nonstructural performance objectives in ASCE 31 and 41 incorporating uncertainty into the definitions.
- Increase programs to move research results into practice and to train engineers and other design engineers with use of latest consensus standards of practice.
- Develop methods to track replacement or retrofit of existing deficient buildings within all occupancy categories, and for each seismic region.
- Develop recommendations for appropriate retrofit of nonstructural systems dependent on local seismicity and occupancy.

- Support updating of standards and guidelines for evaluation, retrofit design methods, and retrofit techniques, and development of new standards and guidelines as appropriate.
- Develop methods to measure the contribution, positive or negative, of the local existing building stock to community resilience.
- Incorporate the concepts of sustainability, including preservation of embedded energy, across all aspects of the existing building seismic issue.
- Encourage risk reduction programs that educate communities and leverage interest in seismic safety and community resilience such as evaluation of schools and community emergency response buildings.

Implementation Issues

Issues associated with the effective implementation of improved techniques of evaluation and retrofit, and more generally, effective reduction of risk in existing buildings, include the following items:

- Lack of awareness of the significance of the risk from older existing buildings with respect to life safety, and perhaps more importantly, as a direct link to the level of community resiliency.
- Lack of training of engineers and other design professionals, particularly architects, planners, and building officials, with the highest priority in areas with less frequent seismic events.
- Lack of confidence regarding cost-effectiveness of current standards and techniques.
- Lack of an integrated program of research and application conditioned by feedback from informed stakeholders.
- The concept of a building rating system that would automatically place value on seismic performance is excellent, but the formalization and implementation of such a system will be difficult. The U.S. Green Building Council and the LEED rating system may provide a useful model.
- The difficulty of building an accurate inventory including the prevalence of specific seismic deficiencies prevents efficient identification of building/structural types to target for replacement or retrofit.

TASK 14: PERFORMANCE-BASED EARTHQUAKE ENGINEERING FOR BUILDINGS

Performance-based earthquake engineering enables decision-makers to target explicit levels of vulnerability for components of the built environment in terms of their resilience (life safety, repair cost and business interruption). Advances in performance-based earthquake engineering will facilitate the development of design tools, codes and standards of practice for new and retrofit buildings, lifelines, and geo-structures; a building rating system; regional loss estimation; structure-specific loss estimation; earthquake-oriented decision tools for individual owners, communities, businesses, and governments; and portfolio analysis.

Proposed Actions

➤ Advance performance-based earthquake engineering to improve design practice, inform decision-makers, and revise codes and standards for buildings, lifelines, and geo-structures.

Existing Knowledge and Current Capabilities

Design of buildings and other structures for earthquake shaking was developed initially to avoid collapse and to prevent debris from falling into adjacent streets. The first design rules were based on estimates of the required lateral strength based on observation of the performance of structures in damaging earthquakes in Japan, Italy, and the United States but had little or no scientific basis. Continuing observation of the performance of structures under strong ground shaking, as well as a gradual increase in understanding of dynamic structural response to shaking, led to refinement of these design rules over the past 6 decades. Over that time, the performance goal of seismic design remained as "life safety" although the term was only vaguely defined. Structures deemed important, such as nuclear power plants, key bridges, and post-earthquake emergency buildings were designed with the intention of controlling or minimizing damage, generally by making them stronger. However, the seismic design of most buildings and structures today relies on design rules rather than analysis of the structure under expected shaking to estimate damage.

Many older pre-code buildings and structures were known to be high-risk but the design rules for new buildings and other structures were difficult or impossible to apply to reduce this risk. New rules were created for this purpose, often for a specific structure type (e.g., unreinforced masonry bearing wall buildings) and it was acknowledged that the seismic performance of these retrofitted buildings and structures would not be equal to that of new construction. As retrofitting became more common, trade-offs between cost and disruption of the work and expected performance was accepted as an inevitable characteristic of earthquake-risk reduction. When the FEMA-funded project to develop formal engineering guidelines for retrofit of existing buildings began in 1989 (ATC, 1994) it was recommended that the rules and guidelines be sufficiently flexible to accommodate a wide range of local, or even building-specific, seismic risk reduction policies. The resulting document, FEMA 273, *NEHRP Guidelines for the Seismic Rehabilitation of Buildings* (FEMA, 1997a), contained various performance levels with titles of Operational, Immediate Occupancy, Life Safety, and Collapse Prevention and the seismic design of buildings for desired performance, rather than to prescriptive rules, began to gain traction in the design professional community. Because of these developments, this focus area is strongly related to Task 13 above.

Increases in analytical capability and a growing demand for performance-targeted seismic design led to a FEMA-funded program to develop guidelines for performance-based design of buildings, which is several years from completion and is operating under a budget significantly reduced from that initially determined to be required. The ATC-58 project focuses on the calculation of repair cost, time of business interruption, and likely casualties in the building due to earthquake shaking. There are great uncertainties related to all aspects of these calculations, including what intensity of ground motion is expected; the exact characteristics of the ground motions; the accuracy of the computer model and the analysis method; the nature of the damage to the structural framing, nonstructural components, and building contents; and the consequences of this damage. The Guidelines will explicitly consider all these uncertainties resulting in a relatively complex methodology that will have to be simplified for practical use. A 50 percent draft of the Guideline (ATC, 2009b) is available from the Applied Technology Council.[27]

Geo-structures, which are engineered earthen construction and include levees, dams, and landfills, are critical components of the built environment. Performance-based design and assessment tools for this important class of structures are unavailable.

Herein, buildings and other structures as defined by FEMA P-750, *NEHRP Recommended Provisions for the Seismic Design of New Buildings and Other Structures* (FEMA, 2009b), and industrial and power-generation facilities are described as structures, lifelines include bridges and transportation networks, and geo-structures are engineered earthen construction as noted above. Earthquake-induced ground deformation is assumed to include surface fault rupture, landslides, liquefaction, lateral spreading, and settlement.

Enabling Requirements

The basic knowledge needs and implementation tools required to advance performance-based earthquake engineering include the follow-

---

[27] See www.atcouncil.org.

ing items, many of which are identified in *Research Required to Support Full Implementation of Performance-Based Seismic Design* (NIST, 2009).

- ANSS should be fully deployed and maintained (see Task 2) to improve knowledge of seismic wave propagation, basin effects, local soil effects, ground motion incoherency, effects of embedment, wave scattering, ground deformation, and soil-foundation-structure interaction.
- Earthquake generation and propagation through heterogeneous underground structures is a very complex process that is impossible to observe. Satellite and LiDAR imaging of fault traces, zones susceptible to liquefaction and landslides, and paleoseismic studies are needed to better characterize earthquake sources and impacts.
- Performance-based seismic design and assessment use—and will continue to use—the results of seismic hazard analysis, which characterizes the effects of earthquake shaking. USGS should continue to update the National Seismic Hazard Maps and the associated design-oriented Java-based applets using new knowledge on earthquake ground motion, faults, and predictive relationships.
- Predictive models of ground shaking and deformation are required for performance-based earthquake engineering. In those regions of our nation where earthquake data are sparse or nonexistent, earthquake-physics simulations should be used to build or augment the dataset. USGS should develop urban hazard maps (where "urban" extends beyond coastal California; see Task 4) for liquefaction (including lateral spreading and settlement), surface fault rupture, and landslide potential, to complement maps available for ground shaking.
- Excessive ground deformation can fail foundations, lifelines, and geo-structures. Robust analysis procedures for predicting ground deformation and their effect on elements of the built environment are needed. Techniques for mitigating the effects of liquefaction should be developed and validated using NSF-NEES infrastructure.
- Tools for site response analysis range in complexity between deterministic site class coefficients and nonlinear site response analysis. The deterministic site class coefficients that are provided in ASCE 7 (ASCE, 2005) are approximate values, are based on limited data, and are strictly applicable to West Coast sites only. Nonlinear site response analysis uses constitutive models for soils, which vary in rigor and degree of validation, especially for deformations consistent with the intense shaking expected close to active faults. Improved site class coefficients, which are valid for sites across the United States, are needed for routine design and performance assessment. Improved constitutive models for soils are required to enable robust nonlinear site response analysis.
- Soil-foundation-structure interaction can substantially modify the

seismic response of a structure. Advanced time- and frequency domain simulation algorithms, codes and tools for such analysis (1-dimensional, 2-dimensional, and 3-dimensional) of discrete structures and clusters of structures (dense urban region) are needed for accurate assessment of performance. Improved constitutive models for soils will enable these simulations. Simplified guidelines and tools to address soil-foundation-structure interaction for scheme design and seismic performance assessment must be developed.

• Performance-based seismic design and assessment of structures typically involve a suite of three-component sets of earthquake acceleration time series. There is no consensus on how to select and scale these time series, and no single method will be widely applicable. The optimal procedures may vary as a function of earthquake intensity, geographic location (seismic hazard), local soil conditions, dynamic properties of the structure, and proximity to adjacent structures. Reliable procedures to select and scale earthquake ground motions for design and assessment of structures, lifelines, and earthen structures must be developed.

• Performance and loss computations are made by analysis of a numerical model(s). Numerical models of structures are assemblies of models of structural and nonstructural components. Improved hysteretic models are required for modern and archaic structural components using physics-based constitutive micro models and data from tests of large-size components and systems using NSF-NEES infrastructure. Component models should be able to trace behavior under arbitrary loading through to failure.

• Current procedures for performance-based earthquake engineering, such as those included in ASCE 41-06, *Seismic Rehabilitation of Existing Buildings* (ASCE, 2007), have not been benchmarked adequately. Many engineers opine that the procedures are conservative and that their use leads to needless construction expenditure, which impedes voluntary seismic rehabilitation. The reliability of the ASCE 41 procedures is unknown. A systematic examination of the procedures, using earthquake data and other evaluation methodologies, is needed because these procedures will not be replaced in the near term.

• Codes and standards of seismic design practice target the prevention of collapse (structures, lifelines) or catastrophic failure (earthen structures). Collapse or failure may result in catastrophic financial and/or physical losses. Current methods for collapse calculations are unproven and likely unreliable because the mechanisms that trigger collapse or failure are not understood and component and constitutive material models cannot trace behavior through failure. Substantial improvements in numerical modeling tools are needed for collapse computations. These tools must be validated by a combination of small-scale and full-scale physical simulations of systems through failure using NSF-NEES infrastructure.

•   Loss computations use fragility and consequence functions for modern and archaic structural and nonstructural components and assemblies in structures. The database of such functions for components and assemblies is small and must be expanded through coordinated numerical and experimental simulations using NSF-NEES infrastructure and collaboration between researchers and design professionals.

•   The loss estimation tools developed by FEMA for the ATC-58 project on performance-based seismic design of buildings are basic (ATC, 2009b). Loss estimation tools must be advanced for structures and expanded to address loss associated with ground deformation, fire following earthquake, and carbon emissions associated with earthquake-induced damage. A recent study by NIST developed a research agenda to fully realize the benefits of performance-based seismic design as envisaged by the ATC-58 project (NIST, 2009).

•   The expected performance of code-conforming structures across the range of intensity of earthquake shaking is unknown. Performance assessment tools such as ASCE 41-06 and the draft ATC-58 methodology should be used to assess the likely performance of modern, code-conforming buildings as a function of framing-system type and height, local soil conditions, and geographic location (seismic hazard) and to inform future revisions of design codes and standards.

•   The performance-based earthquake-engineering framework developed in the ATC-58 project should be extended to address the effects of ground deformation and flooding, and expanded in scope to enable design and assessment of non-building structures including lifelines, earthen structures, and flood protection systems.

•   Nonstructural components and contents comprise most of the investment in structures but there are no performance-based seismic design procedures for such components and contents. Such procedures should be developed but be informed by work completed over the past 2 decades by the U.S. Nuclear Regulatory Commission.

•   Construction materials and framing systems are by-and-large unchanged from those used 50 years ago. Smart/innovative/adaptive/sustainable structural framing systems provide new opportunities for construction and warrant speedy development.

•   A method should be developed to directly calculate the carbon footprint of proposed buildings and the potential savings available from providing better seismic performance.

•   Performance-based seismic design and assessment is computationally more intensive than traditional code-based design. Each simulation may generate gigabytes of data. New visualization tools are needed to assess these large datasets. Grid- and cloud-based computing tools will be required to support these large-scale simulations.

• A plan should be developed and executed to regularly revise guidelines, standards, and codes for performance-based design and assessment of buildings.

Implementation Issues

Issues associated with the effective implementation of performance-based earthquake engineering include the following items.

• Guidelines, standards, and codes are the primary mechanism for improving design practice. Federal agencies should develop and execute a plan to regularly revise guidelines, standards, and codes for new and retrofit design and performance (loss) assessment of facilities and infrastructure.

• The web is now the preferred portal for earthquake-related products and data. Web-based products and earthquake-related data should be further developed and maintained to transfer new knowledge to end users.

• Acceptance by stakeholders of explicit inclusion of uncertainty in descriptions of expected losses.

• Development of a simplified procedures and tools appropriate for use by design professionals for routine design.

• Development of educational materials for owners of buildings that will enable and encourage use of performance-based design.

• Educational programs for design professionals (see Task 17).

• Calibration of design rules in building codes to achieve a specified level of performance in design earthquake shaking.

TASK 15: GUIDELINES FOR EARTHQUAKE-RESILIENT LIFELINE SYSTEMS

Reliable infrastructure is a priority goal for earthquake-resilient communities. The capacity for critical infrastructure to withstand and quickly restore services following an earthquake or other natural or man-made disasters determines how rapidly communities can recover from such disasters. Many communities rank the availability of electricity, highways, and water supply as the top three critical infrastructure or lifeline systems that need to function following an earthquake (ATC, 1991). Reliable infrastructure is also recognized as being essential at the national level for global economic competitiveness, energy independence, and environmental protection (NRC, 2009). Reducing the vulnerability of critical infrastructure to natural disasters is identified as a strategic priority in the 2008 NEHRP Strategic Plan, by the NEHRP Advisory Committees (ACEHR, 2009), and in congressional testimony (O'Rourke, 2009) and is linked to broader federal policies and priorities, including the Department of Homeland Security's *National Infrastructure Protection Plan* (DHS, 2009) and the Subcommittee on Disaster Reduction's *Grand Challenges for Disaster Reduction* (SDR, 2005).

Although some infrastructure renewal is being addressed in the short term through the American Recovery and Reinvestment Act (PL 111-5), there is much to do in the long term, as documented in reports such as the American Society of Civil Engineers' Infrastructure Report Card (ASCE, 2009), which gave 12 of America's infrastructure categories an overall grade of D and estimated an investment of $2.2 trillion over 5 years to upgrade and improve infrastructure systems. America's infrastructure is made all the more vulnerable to earthquakes and other natural disasters by its poor state of repair. Large segments of the nation's critical infrastructure are now 50 to 100 or more years old, and many were built long before the current generation of earthquake codes, standards, and guidelines were put into effect. In California, past earthquakes have helped to identify and damage the weak links in infrastructure systems, and many owners have generally taken steps to adequately prepare for future disasters through repair and replacement programs and the implementation of improved standards and guidelines, updated construction materials, and current design practices. Yet there is still much to do. The catastrophic levee failures in New Orleans following Hurricane Katrina in 2005 demonstrated the vulnerability of these lifeline systems. A strong San Francisco Bay area or Central Valley earthquake could result in failure of the levee system in the Sacramento-San Joaquin Delta with consequent disruption to drinking water supplies to more than 22 million Californians; disruption of irrigation water to Delta and state agricultural lands could cascade into a national agricultural disaster.[28] Elsewhere in the United States where earthquakes are less frequent, the vulnerabilities or risks posed by earthquakes to infrastructure systems may not be recognized or fully appreciated by stakeholders. As a result many owners may have only partially prepared, or have done nothing at all.

A dramatic "wake up call" concerning the vulnerability of electric systems and the resultant regional and national consequences occurred as a result of the August 2003 Northeast Blackout. This blackout affected 5 states, 50 million people, and caused an estimated $4-10 billion in business interruption losses in the central and eastern United States (U.S.-Canada Power System Outage Task Force, 2004). Moreover, the power

---

[28] See www.water.ca.gov/news/newsreleases/2005/110105deltaearthquake.pdf.

outage caused "cascading" failures to water systems, transportation, hospitals, and numerous other critical infrastructures; such infrastructure failure interdependencies are common across many types of disasters (McDaniels et al., 2008). In 1998, a study on the effects of a large New Madrid earthquake in the central United States estimated that direct and indirect business interruption economic losses due to extended power disruption could be as high as $3 billion (Shinozuka et al., 1998). At that time, there was little evidence that such losses were possible. The 2003 Northeast Blackout demonstrated that while initiating events can vary (e.g., a falling tree, an earthquake, or an act of terrorism), the consequences can be similar.

Proposed Actions

The 2008 NEHRP Strategic Plan identified the development of earthquake-resilient lifeline components and systems as a Strategic Priority. In contrast to individual buildings, which occupy a specific site or location, lifeline systems are geographically distributed and interconnected. As a result, these systems have earth science and engineering needs that may be specific or unique to a particular lifeline system and differ from those of the building community. For example, geo-structures are engineered earthen construction and include levees, dams, and landfills. As discussed in Task 14, these critical components of the built environment currently lack performance-based design and assessment guidelines. Guidelines, standards, and codes are the primary mechanism for operating and maintaining functional infrastructure systems. Typically, standards and guidelines address individual components of lifeline systems, and while many contain earthquake loading and design provisions, others do not contain such provisions. The actions that are needed to make lifeline systems more resilient are:

1.   Fill in critical remaining gaps. New standards and guidelines that fill in the remaining gaps for lifeline performance and retrofit should be developed during the initial 5-year period. The American Lifelines Alliance Matrix of Standards and Guidelines for Natural Hazards (ALA, 2003) summarized the natural and man-made hazard provisions of infrastructure standards and guidelines, and this summary provides a framework for identifying where guidance needs to be developed, improved, or updated. As discussed in Task 10, there is a need to better characterize infrastructure network vulnerability and resilience. This would identify the weaknesses in current lifeline systems and the consequences of lifeline interdependencies (both spatial and functional) in order to prioritize the most effective retrofits and functional modifications to improve future earthquake performance at the regional and community level.

There are few guidelines/standards that address system reliability, i.e., the practices that are developed to provide reasonable assurance that the consequences of a natural hazard event on system service will meet the goals established by the stakeholders. Like performance-based engineering for buildings, these consequences are defined by multiple requirements, but typically include public safety, duration of service interruption, and the costs to repair damage. Tools are needed to model the effects of these consequences not only to the utilities, but also on the local community and economy (see Task 10).

2. Systematically review and update existing lifelines standards and guidelines. Existing lifelines standards and guidelines need to be systematically reviewed and updated to include the most up-to-date utility practices and the latest engineering and geotechnical research results. The need for a consensus on the level of hazard that should be considered for use in new lifeline designs and upgrades was articulated at the community workshop hosted by this committee. NEHRP collaboration with Standards Developing Organizations can facilitate these types of reviews and coordinate code and standard updates that reference the latest edition of the National Seismic Hazard Maps.

3. Demonstration/pilot projects tied into pilot communities. Reliable electric power and water are essential for developing community resiliency. Pilot programs and demonstration projects that showcase new utility practices and implement lifeline mitigation guidelines and standards should be encouraged as part of the larger community pilot NEHRP programs discussed in Tasks 17 and 18. Involving community stakeholders in defining the level of acceptable lifeline risk for their communities, and understanding shared public-private responsibilities, are necessary to achieve those goals.

4. Lifelines earthquake engineering research. Lifelines-focused research is needed to fill many of the gaps identified in Action 1 above. NEHRP-supported collaborative research, with infrastructure owners and operators, to address user- and owner-defined issues has been a success in the past and should be reinvigorated during the next 5 years. The community workshop in Irvine, CA, identified a number of lifeline research topics that cut across engineering, earth science, and social science issues, including the need to better understand lifeline interdependencies and the physical performance of lifeline systems and the consequences to communities of their disruption, and the need to develop protocols for researchers to use proprietary data for analysis without jeopardizing security and confidentiality of the lifeline system operator. The workshop

also recognized that active support for this type of research from both the public and private sectors is a key requirement for continued success.

Existing Knowledge and Current Capabilities

Critical infrastructure or lifeline systems are the utility—energy (electric power, natural gas, and liquid fuels), water, wastewater, and telecommunications—and transportation (highway, rail, water ways, ports and harbors, and air) systems (NRC, 2009). The ownership and responsibility for critical infrastructure systems span both public and private sectors. Water and waste water systems are primarily owned and operated by public entities, while the private sector typically owns and operates power and telecommunications systems. State and local authorities are responsible for roads, highways, and bridges; and ports, airports, railroads are owned by quasi-public or private organizations. Regulatory oversight for infrastructure systems is equally broad and spans federal, state, and local jurisdictions. In contrast to individual buildings, which occupy a specific site or location, infrastructure systems are geographically distributed and interconnected. These types of networks create functional and geographic interdependencies, where damage in one part of the system can impact other parts of system, and damage in one lifeline system can disrupt other systems. Many interdependencies may have been created by default, not by plan, creating unforeseen vulnerabilities that are not apparent until a disaster strikes.

Lifeline Earthquake Engineering Research

The 1906 San Francisco and 1933 Long Beach, CA, earthquakes demonstrated the consequences of multiple lifeline systems failures on a community. Although the severe effects of these earthquakes spurred the initial development of seismic design requirements in buildings and other structures in California, lessons on the need for rapid restoration of lifelines to aid in community response waned as a result of the lack of a significantly damaging urban earthquake for nearly 4 decades after 1933.

The 1971 San Fernando, CA, earthquake is regarded as "birth of lifeline earthquake engineering" in the United States. The catastrophic effects from this earthquake to infrastructure in a rapidly growing area of southern California stimulated efforts in the engineering community to address the vulnerabilities that were exposed. ASCE created the Technical Council on Lifeline Earthquake Engineering (TCLEE) in 1974 to advance the state of the art and practice in lifeline earthquake engineering through research, standards and guideline development, and implementation at operating utility systems. TCLEE has actively published a monograph series on lifeline topics and conducts post-earthquake reconnaissance surveys of lifeline performance after major earthquakes in the United States and worldwide.[29]

In the decades following the San Fernando earthquake, NSF-sponsored Engineering Research Centers (National Center for Earthquake Engineering Research, now the Multidisciplinary Center for Earthquake Engineering Research (MCEER) at the State University of New York, Buffalo;[30] the Mid-America Earthquake Center (MAE) at the University of Illinois at Urbana-Champaign;[31] and the Pacific Earthquake Engineering Research Center (PEER) at the University of California, Berkeley[32]) were established to carry out research with the goal to reduce losses due to earthquakes. Each center developed a specific focus for its research and development activities. The central focus of PEER, for example, was performance-based earthquake engineering, facility- and system-level models, and computational tools for assessing and reducing earthquake impacts. MAE, on the other hand, focused on consequence-based engineering, system-level simulation, and analysis for assessing and reducing impacts. MCEER focused on the use of advanced and emerging technologies for reducing impacts and developing methods to quantify community resilience. All three earthquake engineering centers are also participants in the NSF-funded NEES program.[33] NSF funding for these engineering research centers has now ceased, and the level of center-based research on lifelines has decreased substantially. This decrease impacts not only engineering research, but also interdisciplinary research as well.

Performance-Oriented Standards and Guidelines

The NEHRP agencies also addressed lifeline issues during the 1980s and 1990s through a series of workshops and studies. A nationwide assessment of lifeline seismic vulnerability and the impact of lifeline system disruption was conducted by the Applied Technology Council (ATC, 1991), which ranked the electric system, highways, and the water system as the most critical lifelines in terms of the impact of damage and disruption. Workshops conducted by NIST and FEMA (e.g., NIBS, 1989; FEMA, 1995; NIST, 1996) noted the limited number of nationally recognized standards for the design and construction of lifeline systems at that time, and recommended a focus on system performance as well as component performance consistent with the hazard level in the region. In 1997, an ASCE

---

[29] See www.asce.org/community/disasterreduction/tclee_home.cfm.
[30] See mceer.buffalo.edu.
[31] See mae.cee.uiuc.edu.
[32] See peer.berkeley.edu.
[33] See www.nees.org/.

Lifeline Policy Makers Workshop (NIST, 1997) recommended emphasis on guideline development and demonstration projects. Development and implementation of those recommendations were estimated to cost $15 million over 5 years.

In 1998, a cooperative agreement between ASCE and FEMA led to the creation of the ALA. ALA's objectives were to facilitate the creation, adoption, and implementation of national consensus standards and guidelines to improve lifeline performance during hazard events. The ALA strategy focused on using the best industry practices, involving Standards Development Organizations (SDO), and addressing *both* component and network performance. In late 2002, FEMA brought the ALA under the Multi-hazard Mitigation Council through a partnership with NIBS. The ALA developed the Existing Guidelines Matrix (see Table 1 in ALA, 2003) to summarize the current status of natural and man-made hazards guidance available in the United States for lifeline system operators. Lifeline design and assessment guidelines and standards prepared by SDOs, professional and industry organizations, and practitioners in the relevant fields were included to identify the needs for guidance that does not exist yet or that must be improved and updated. ALA activities were terminated in 2006 because of budget restrictions at FEMA, severely hampering efforts toward further improving national lifeline standards and guidelines.

In addition to NEHRP support for lifeline research and guideline development, other federal and state agencies (e.g., Department of Energy (DOE), Federal Highway Administration (FHWA), Department of Transportation (DOT)) and the private sector (Electric Power Research Institute (EPRI), consortium projects with PEER/MCEER/MAE, Cooperative Research and Development Agreement (CRADA) projects with USGS and other federal agencies) actively support lifelines research. Cooperative user-driven research in the MCEER and PEER Lifelines programs, for example, has brought the state (CalTrans, CA Energy Commission) and private sector (Los Angeles Department of Water and Power (LADWP), Pacific Gas and Electric Company (PG&E), and other utilities) together with researchers to address common interest topics.

Enabling Requirements

Research supported by NEHRP has substantially improved the modeling of complex lifeline systems, structural health monitoring, protective systems for buildings and bridges, and remote sensing for response and recovery from extreme events (EERI, 2008). NSF support for NEES has provided a national resource for demonstrating the cost-effectiveness of performance-based design, the development of new materials to reduce the impact of earthquakes and other extreme events, and the creation of retrofit strategies to improve existing infrastructure performance. A continued commitment to improve performance-based design and engineering practices, coupled with development of physics-based numerical simulations for both component and system performance, is necessary for next generation of earthquake-resilient lifeline systems. The systematic documentation and archiving of lifeline earthquake performance data (see discussion of PIMS, Task 9) is essential to evaluate these types of simulation models. Engineering objectives that address entire lifeline system performance (e.g., outage targets for extreme conditions) need to be developed for local needs and conditions.

In addition to mapping geologic hazards along lifeline corridors, geotechnical research to improve strong ground motion (wave passage, spatial coherency, and duration effects), ground displacement/deformation (fault rupture, landslides, liquefaction), and damage estimates is needed to more accurately describe the earthquake demands on lifeline system. Social and economic research to better understand the societal consequences of lifelines failures on communities is also needed. Emergency protocols for response to catastrophic lifeline failures, such as dam failures or natural gas-fueled fires following an earthquake, need to be reviewed as communities grow and encroach on potentially hazardous areas. The ability to model cascading failures of, and between, social and infrastructure systems can help a community visualize the impacts and identify the necessary steps to become more resilient.

Federal coordination of earthquake-related lifeline research and mitigation, however, needs to extend beyond the four principal NEHRP agencies. The 2008 NEHRP Strategic Plan states that it will "focus its efforts on critical lifelines systems and components that are not being addressed by other agencies or organizations, in order to avoid duplicative efforts and maximizing leveraging of resources." This goal recognizes the need to bring other federal agencies that either support research or have regulatory authority, such as the Department of Energy, the Department of Transportation/Federal Highway Administration and the Office of Pipeline Safety, and the Department of Homeland Security "to the table" in order to leverage investments and optimize potential NEHRP contributions. In addition to federal coordination, multi-level coordination between all stakeholders at state and local levels (including both public and private sectors) is critical for successful lifeline risk management.

Implementation Issues

Utilities are familiar with preparing for and responding to natural hazard events such as strong windstorms and seasonal floods or human events such intentional acts of vandalism or accidental "dig ins" or rupturing of buried pipelines. Rare and extreme events, such as severe earthquakes, flooding of historic proportions, or a concerted terrorist attack, however, can overwhelm ordinary utility experience and preparation and can result in widespread damage and service disruption. Although utilities that have experienced such severe events have generally taken steps to adequately prepare for future disasters, many others have only partially prepared, and still others are not aware of their full exposure or vulnerability to these threats. The current lack of an organization like the ALA to facilitate the creation, adoption, and implementation of national consensus guidelines and standards to improve lifeline performance during rare or extreme hazard events is a major impediment to implementation of NEHRP goals.

Investment Priorities

Various stakeholders, both public and private, have competing priorities for risk management investments. In some cases those investments can be at the expense of or delay seismic mitigation activities such as equipment or building retrofits, especially in areas of perceived low seismic hazard or risk. An unintended consequence of the restructuring of the electricity industry in the United States, for example, has been a sharp decline in expenditures for research and development by investor-owned utilities (Blumstein and Wiel, 1999).

Confidentiality Issues

Protocols for dealing with proprietary data and analysis, without jeopardizing the security and confidentiality of the lifeline system operator, need to be developed at federal, state, and local levels. Many stakeholders, especially those in areas of critical infrastructure, are reluctant or prohibited to release inventory information outside of their organizations. These restrictions impact the ability of communities to recognize and plan for service disruptions during disasters. Public-private partnerships, whereby individual utilities could conduct their own risk assessments, using standardized methodologies and earthquake scenarios, inside the company and then share the results with their counterparts and other stakeholders to address inter-utility interdependencies and community impacts need to be encouraged. These types of partnerships would allow for more informed disaster planning within the community without sacrificing confidentiality issues.

TASK 16: NEXT GENERATION SUSTAINABLE MATERIALS, COMPONENTS, AND SYSTEMS

The construction materials used in seismic framing systems in medium- and high-rise buildings and other structures are either concrete or steel, and both of these materials have a high carbon footprint. There have been few materials developments in the past 100 years. New sustainable materials suitable for use in the construction industry should be developed to meet the goals of high performance (and thus low volume) and low carbon footprint per unit volume. Buildings constructed with these components should enhance the earthquake resilience of the built environment.

Adaptive components and framing systems have been proposed in the form of semi-active and actively controlled components and structures but have not been implemented in buildings and other structures in the United States. Adaptive components offer the promise of better controlling the response of structures across a wide range of shaking intensity to limit damage and loss.

Proposed Actions

➤ Develop and deploy new high-performance materials, components, and framing systems that are green and/or adaptive.

Existing Knowledge and Current Capabilities

Little research and development effort has been devoted in the past 3 decades to new materials for application to earthquake-resistant construction. Notable exceptions include fiber-reinforced polymers for retrofit applications and elastomers and composites in seismic isolation systems. Some work is under way on low-cement concretes, fiber-reinforced high-performance concretes, and very high-strength steel. Despite these innovations, the field application of these emerging technologies is stymied for a number of reasons, including (a) incomplete materials characterization, (b) high perceived cost, (c) lack of regulation and/or design standards, (d) a conservative and risk-averse construction industry, and (e) limited incentives for green construction.

Structural components constructed using adaptive fluids (e.g., electro- and magneto-rheological fluids) and bracing systems have been tested in the laboratory at small and moderate scales (e.g., Whittaker and Krumme, 1993; Spencer and Soong, 1999; Soong et al., 2005). The advantages offered by adaptive components have been explored but not documented, with the advantages being dependent on the control algorithms used, and also the need for external power sources for actuating the components and powering the sensors. There is no guidance or standards for implementing adaptive components in buildings and other structures, and there are no suppliers of adaptive products suitable for implementing in buildings and structures.

Enabling Requirements

The basic knowledge needs and implementation tools required to develop and deploy high-performance, sustainable, and/or adaptive materials and framing systems for earthquake-resistant construction include the following items. Herein, buildings and other structures as defined by FEMA P-750, *NEHRP Recommended Provisions for the Seismic Design of New Buildings and Other Structures* (FEMA, 2009b), are denoted as structures.

- Investigate and characterize new materials, including but not limited to (a) low-cement concrete, (b) cement-less concrete, (c) very high-strength concrete, (d) steel- and carbon-fiber-reinforced concrete, (e) very high-strength steel, and (f) fiber-reinforced polymers. Characterize new materials across a wide range of strain, strain rate, temperature (including fire), and environmental exposure.
- Devise new modular pre-cast components and framing systems that best utilize the new materials, such as sandwich construction involving permanent steel shells that function as formwork and reinforcement and infill low-cement (or cement-free) concretes.
- Develop tools, technology, and details to join components constructed with new materials.
- Prototype components, connections, and framing systems.
- Conduct moderate-scale and full-scale tests of components constructed with new materials using NEES infrastructure to characterize component response in sufficient detail to enable the development of design equations suitable for inclusion in a materials standard, hysteretic models for nonlinear response analysis, and fragility functions for performance-based seismic design and assessment.
- Conduct near full-scale tests of complete three-dimensional framing systems constructed using new materials and/or components using NEES infrastructure and/or the E-Defense[34] earthquake simulator.
- Develop design tools and equations for each new material, component, and framing system and prepare a materials standard similar in scope to ACI 318 (ACI, 2008). Actively support the standard-development process, its implementation in the model building codes, and its adop-

---

[34] See www.bosai.go.jp/hyogo/ehyogo/.

tion by the design professional community. Assign seismic parameters for routine code-based design using established procedures such as those presented in FEMA P-695, *Quantification of Building Seismic Performance Factors* (FEMA, 2009a).

•   Prepare consequence functions for components and framing systems constructed with new materials in support of performance-based seismic design and assessment. Use NEES infrastructure for this task and ensure close collaboration between researchers and design professionals.

•   Develop a family of adaptive materials suitable for implementation in structural components, including controllable fluids and shape-memory materials. Characterize new materials across a wide range of strain, strain rate, temperature (including fire), and environmental exposure.

•   Develop a family of robust algorithms suitable for controlling the response of adaptive fluids and metals and traditional structural components.

•   Develop a family of low-cost, low-power, zero maintenance wireless sensors suitable for controlling the response of adaptive components and monitoring the health and response of structural framing systems.

•   Prototype adaptive components (devices, materials, and sensors).

•   Develop a suite of algorithms for the control of linear and nonlinear structural framing systems subjected to three components of earthquake ground motion.

•   Conduct moderate-scale and full-scale tests of adaptive components using NEES infrastructure to characterize component response in sufficient detail to enable the development of design equations suitable for inclusion in guidelines and standards, hysteretic models for nonlinear response analysis, and fragility functions for performance-based seismic design and assessment.

•   Conduct near full-scale tests of complete three-dimensional framing systems constructed using adaptive components using NEES infrastructure and/or the E-Defense earthquake simulator.

•   Develop design tools and equations for each new adaptive material and components constructed using that material.

Implementation Issues

Issues associated with the effective implementation of new materials, components, and framing systems include the following items.

•   Acceptance by design professionals, contractors, and building officials and regulators of new materials, components, and framing systems.

•   Development of educational materials to encourage use of high-performance, low-carbon footprint materials.

- Develop financial incentives for the use of construction materials with a low-carbon footprint.
- Lack of familiarity of design professionals, contractors, and building officials with control algorithms suitable for implementation of adaptive materials, components, and framing systems.
- Lack of familiarity of design professionals, contractors, and building officials with sensing and structural health monitoring technologies.
- Lack of guidelines, codes, and standards for the analysis, design, and implementation of adaptive materials, components, and systems.

TASK 17: KNOWLEDGE, TOOLS, AND TECHNOLOGY TRANSFER TO PUBLIC AND PRIVATE PRACTICE

New knowledge and technology will be developed in many of the other tasks described in this report. Analytic and design tools will be developed. Each task description includes a component on education and technology transfer. This overarching task assures that the knowledge and tools developed in other tasks are quickly put into design practice in both the private and public sectors. Long-term continuing education programs should be encouraged to increase the pool of professionals using state-of-art mitigation techniques.

Proposed Actions

➣ Create a new program responsible for coordinating and encouraging ongoing technology transfer across the NEHRP domain that also builds new initiatives to assure that state-of-the-art mitigation techniques are being deployed across the nation.

Existing Knowledge and Current Capabilities

It is generally acknowledged that technology transfer is seldom adequate, and implementation of effective mitigation strategies and techniques is therefore unnecessarily delayed. The incorporation of mandatory education and outreach components into research projects is sometimes effective, but digestion, coordination, and packaging of research results for efficient practical use is often missing. Notable exceptions are NEHRP's support of development of seismic standards and codes for buildings during the past 30 years and support since 2007 of research synthesis and technology transfer to the design professional community through the NEHRP Consultants Joint Venture. Continued support for these programs is needed. However, despite development of codes and standards, training materials for using these codes and standards, and a pipeline for research synthesis and technology transfer, the state of the practice lags far behind the state of the art.

Use of state-of-the-art knowledge and technology from other areas of study that could improve resilience may lag even further behind. There are no systematic programs to consolidate and transfer research results to practice in many disciplinary areas that contribute to seismic resilience such as geotechnical engineering, seismic protection of infrastructure, use of scenarios and regional loss estimation, emergency response, post-earthquake economic recovery, and public policy.

Seismic safety and community resilience is only one of many issues facing most of the implementation community, including owners of buildings and infrastructure, policy-makers at all levels of government, engineers and planners, and the general public. A consistent education and outreach program will not only raise the quality of the state of the practice, but also keep seismic performance issues "on the table."

Enabling Requirements

NEHRP should maintain and re-emphasize existing programs:

• Fully support development of seismic standards and codes of practice for buildings, bridges, lifelines, and mission-critical infrastructure that include transparent statements regarding expected performance. Advocate for their adoption and enforcement.

• Support and expand the development of research synthesis and technology transfer documents and tools through organizations such as Applied Technology Council (ATC), Consortium of Universities for Research in Earthquake Engineering (CUREE), and Building Seismic Safety Council (BSSC).

• Include education and outreach components in research projects.

• Include a strong and significant education and training program in ongoing initiatives such as the Development of Next Generation Performance Based Engineering, the mitigation of risks from existing buildings, and HAZUS. Enable web-based delivery of products.

NEHRP should initiate a new program center that reviews on-going and completed research, couples and coordinates results in different disciplines, and develops outreach and training documents and courses to maximize effectiveness.

Implementation Issues

The primary barrier to successful implementation of this action is the need to create and fund a new unit within NEHRP to coordinate and initiate technology transfer.

TASK 18: EARTHQUAKE-RESILIENT COMMUNITIES AND REGIONAL DEMONSTRATION PROJECTS

The ultimate goal of NEHRP is to make our citizens, our institutions, and our communities more resilient to the impacts of earthquakes, and to ensure that earthquakes will not disrupt the social, economic, and environmental well-being of our society. For the purposes of this report the definition of a resilient nation is one in which its communities, through mitigation and pre-disaster preparation, develop the adaptive capacity to maintain important community functions and recover quickly when major disasters occur. This task supports this ultimate goal by describing a strategy to apply knowledge initially in a number of "early adopting" communities, which ultimately will create a critical mass to support continued adoption nationwide.

The characteristics of an earthquake resilient community are:

- They recognize earthquake hazards and understand their risks.
- They are protected from hazards in their physical structures and socioeconomic systems.
- They experience minimum disruption to life and economy after a hazard event has occurred.
- They recover quickly and with a minimum of long-term effects.

Governments, at all levels, own part of the earthquake risk and are better able to carry out their responsibilities when people and businesses are earthquake resilient. Private investments in resilience have public benefits. Public safety; reduced individual, business, and government financial losses; community character; housing availability and affordability; neighborhood-serving businesses; and architectural and historic resources; are all community values supported by individual, private investments in earthquake resilience.

Proposed Actions

NEHRP-supported activities would support and guide community-based earthquake resiliency pilot projects that apply NEHRP-generated and other knowledge to improve awareness, reduce risk, and improve emergency preparedness and recovery capacity. A strategy—based on diffusion theory—would guide the selection of early-adopter communities and employ diffusion processes tailored for each community to create a critical mass of people and organizations taking appropriate actions within each community and between communities. Demonstration projects would be used to focus attention and to demonstrate the value and feasibility of resilience-enhancing measures.

Existing Knowledge and Current Capabilities

Most implementation programs do not understand, and therefore neglect, the process necessary for individuals and organizations to adopt new policies and practices. Providing documents and information, although absolutely necessary, is not enough. NEHRP should develop a comprehensive strategy—from concept to practice—that addresses the people at the community and regional levels who are responsible for earthquake risk and the ensuing consequences. This will require innovations—ideas, practices, or objects that are perceived as new by an individual or local unit that can adopt them. The diffusion of innovations is the process by which an innovation is communicated through certain channels over time among members of a social system. Diffusion is a process that depends on decisions by individuals or organizations to adopt an idea. Rogers (2003) refers to the decision to adopt an idea as the innovation-decision process, which consists of five steps: (1) knowledge, (2) persuasion, (3) decision, (4) implementation, and (5) confirmation. Better understanding of how potential adopters move through these stages can greatly improve earthquake safety efforts. The following are important diffusion principles:

1.   Mass media channels are effective in creating knowledge of innovations, but inter-personal communication from a "near peer" is needed to decide to adopt an innovation and to change behavior.
2.   More than just the demonstration of an innovation's benefits is needed for the adoption of that innovation.
3.   Characteristics of innovations that affect rate of adoption:

  • Relative advantage—is it better than the current alternative or way of doing things?
      • Compatibility—is it compatible with existing values?
      • Complexity—is it easy to use and/or understand?
      • Validation—can it be tested on a partial basis before adopting?
      • Observability—how easy is it to observe the benefits?

Because diffusion is a socially driven process, people are critical to the spread of new ideas. Diffusion theory provides important insight into the types of people that influence the innovation-decision process and move it forward, thereby hastening the spread and adoption of new ideas. Rogers (2003) divides adopters into Innovators, Early Adopters, Early Majority, Late Majority, and Laggards. The people in each adopter category have different traits and different roles in the diffusion process. He also identifies opinion leaders, who are influential people within a system that others respect and listen to, as important in diffusion efforts; if they adopt, others are more likely to. Opinion leaders can accelerate the diffusion process if they are Early Adopters. Gladwell (2000) has similar ideas—there are a few critical people that are necessary for moving an idea from the Early Adopters to the Early Majority, which he terms "Connectors, Mavens, and Salesmen." Other researchers, such as Watts (Thompson, 2008), disagree and argue that ordinary people can perform these functions as well.

Diffusion theory applies to earthquake risk reduction efforts because the main goal is to change people's behavior so they will take actions to reduce their risk, as opposed to doing nothing or taking actions that actually increase their risk. Behavior change is not an engineering problem, and therefore reducing earthquake risk requires theories and methods from other fields. Diffusion theory provides a framework of ideas that explains why earthquake risk reduction projects succeed or fail, and provides instruction describing how to increase the benefits and impact of future projects.

Early adopters are critically important to the diffusion of innovations; for that reason the strategic selection of pilot communities, in which there is targeted, sustained, and direct linkages between research and application through all five states (i.e., knowledge, persuasion, decision, implementation, and confirmation) is essential to achieving earthquake resilience nationwide.

Enabling Requirements

Building a more earthquake-resilient nation should include a robust capacity-building program that is implemented at the community/grassroots level, based on the theory of diffusion. Such a program should initially focus on a minimum of 10 pilot cities, of which at least 5 would be in key earthquake hazard regions of the country. Sufficient knowledge exists to initiate such a program immediately, although new knowledge from research based on this element and other NEHRP activities would improve the program. The program would have several components:

- A data component to develop community-level hazard and risk profiles as well as socio-political-economic data that will be used to assess a baseline of each community's resilience capacity (see also Task 11).
- A research component to document resilience capacity, identify existing examples of resilience capacity, and estimate its cost and broader implications.
- A grass-roots outreach component to focus on establishing the necessary community-level, public-private partnerships of the influential social, economic, and political stakeholders and leaders for capacity building.
- A post-audit component to measure the cost and effectiveness of various resilient actions.
- A demonstration component, perhaps projects to reduce earthquake risk in schools, which would attract attention and demonstrate the value and feasibility of mitigation projects.
- An analysis component to identify gaps between resilience capacity and loss estimation, using different earthquake scenarios.
- An implementation component to work to reduce the gaps and document the results.

Implementation Issues

- A federal entity should be authorized to prepare and carry out a strategy to achieve earthquake resilience at the community and regional levels nationwide.
- Matching grants are needed for approximately 5 years for early-adopter communities to participate.
- The strategy should include measures to sustain implementation efforts over time, and a strategy to increase to a nationwide scale. At a minimum, the strategy should:

    o Begin with a minimum of 10 early-adopter pilot communities to develop techniques for other communities to benefit from and emulate;

    o Develop a nationwide network of community leaders (mavens) interested in earthquake resilience;

    o Involve the private sector as equal and critical partners in the process. Businesses benefit from earthquake-resilient communities in myriad ways commensurate with the nature of their businesses. Businesses that understand the benefits are more apt to invest in their own resilience and offer community leadership, political support, and some incentives;

o Involve more grass-roots-level community organizers who can help disseminate and build interest and support within a community and community-level organizations;

o Require leveraging of resources;

o Incorporate new communication tools including social media;

o Address the vulnerability of buildings and lifelines, social organizations, community values, and government continuity;

o Of necessity, address other hazards that threaten communities.

• Governments should exercise their enforcement powers in ways commensurate with the community and its tolerance for risk: enforcing building codes and land-use restrictions, requiring owners of existing buildings to reduce vulnerability, and encouraging other actions that are generally intended to promote health, safety, and welfare.

• Governments should champion social justice issues raised by variations in vulnerability to earthquake risk; earthquake resilience should not be reserved for those with resources and position.

• The NEHRP implementation program needs to advocate incentives to promote the societal benefits from earthquake risk management practices, and to remove obstacles and disincentives. Meaningful incentives that represent societal value are needed to encourage and reward investments. Incentives are needed to make measures affordable (reduce initial costs and make funds available—loans) and manageable (payable over time), with the return in terms of increased safety and financial security proportional to the investment. Incentives include federal and state tax credits for building owners, accelerated appreciation for businesses, subsidies and grants (matching) for those who provide government-like services (affordable housing, medical clinics and hospitals, schools, etc.) and grants (matching) and eligibility for cost reimbursement for government agencies. Local tax credits, property tax reduction, or transfer tax incentives can exert powerful influence. Mechanisms should also be made available to insurance companies and by insurance companies to increase insurance coverage and encourage mitigation for the earthquake hazard.

• A robust constituency base needs to be developed to advocate on behalf of the entire community. Professional and trade associations should provide leadership in advocacy matters at all levels of government and throughout their respective professional discipline.

• Create partnerships and with the media and recruit them to become early adopters.

4

Costing the Roadmap Elements

The charge for this study required that the committee "estimate program costs, on an annual basis, that will be required to implement the roadmap." The committee was directed to consider the detailed cost estimates presented in the 2003 Earthquake Engineering Research Institute (EERI) report (EERI, 2003b), and validate or revise these estimates. In its deliberations, the committee initially focused on the 2008 NEHRP Strategic Plan, analyzing its goals, objectives, and strategic priorities, and then reviewed the EERI plan and cost estimates. Ultimately, the 18 tasks described in the previous chapter—the elements of the roadmap—are far broader in scope than the elements of the EERI plan, and consequently the costing estimates presented here are substantially different from those that were presented in EERI (2003b).

In estimating costs to implement the roadmap, the committee recognized the high degree of variability among the 18 tasks—some (e.g., deployment of the Advanced National Seismic System [ANSS]) are well developed and actually in the process of being implemented, whereas others are only at the conceptual stage. Costing each task required a thorough analysis to determine scope, implementation steps, and linkages or overlaps with other tasks. For some of the tasks, the necessary analysis had already been completed in workshops or other venues, and realistic cost estimates were available as input to the committee. For other tasks, the committee had nothing more to go on that its own expert opinion, in which case implementing the task may require some degree of additional detailed analysis.

Table 4.1 lists the cost estimates for each task for implementation

TABLE 4.1 Compilation of Cost Estimates by Task, in Millions of Dollars[a]

Task	Annualized Costs (av.) Years 1-5 ($)	Total Cost Years 1-5 ($)	Total Cost Years 6-20 ($)	Total Cost ($)
1. Physics of Earthquake Processes	27	135	450	585
2. Advanced National Seismic System (ANSS)[b]	66.8	334	1,002	1,336
3. Earthquake Early Warning	20.6	103	180	283
4. National Seismic Hazard Model	50.1	250.5	696	946.5
5. Operational Earthquake Forecasting	5	25	60	85
6. Earthquake Scenarios	10	50	150	200
7. Earthquake Risk Assessments and Applications	5	25	75	100
8. Post-earthquake Social Science Response and Recovery Research	2.3	11.5	TBD[c]	TBD[c]
9. Post-earthquake Information Management	1	4.8	9.8	14.6
10. Socioeconomic Research on Hazard Mitigation and Recovery	3	15	45	60
11. Observatory Network on Community Resilience and Vulnerability	2.9	14.5	42.8	57.3
12. Physics-based Simulations of Earthquake Damage and Loss	6	30	90	120
13. Techniques for Evaluation and Retrofit of Existing Buildings	22.9	114.5	429.1	543.6
14. Performance-based Earthquake Engineering for Buildings	46.7	233.7	657.8	891.5
15. Guidelines for Earthquake-Resilient Lifelines Systems	5	25	75	100
16. Next Generation Sustainable Materials, Components, and Systems	8.2	40.8	293.6	334.4
17. Knowledge, Tools, and Technology Transfer to Public and Private Practice	8.4	42	126	168
18. Earthquake-Resilient Communities and Regional Demonstration Projects	15.6	78	923	1,001
TOTAL	**306.5**	**1,532.3**	**5,305.1**	**6,837.4**

[a] See following section for explanatory notes (all figures are 2009 dollars).

[b] Does not include support for geodetic monitoring or geodetic networks.

[c] Funding during the remaining 15 years of the plan would be based on a performance review after 5 years.

time-frames of 0-5 years, 6-20 years, and the overall 20-year total. In summary, the annualized cost for the first 5 years of the program for national earthquake resilience is $306.5 million/year.

EXPLANATORY NOTES FOR COSTING

Much of the finer detail used as the basis for task costing is presented in Appendix E. The following is summary information (using 2009$) to assist with reading the cost estimates presented in Table 4.1.

Task 1—Physics of Earthquake Processes

Basic research on the physics of earthquake processes is supported by the National Science Foundation (NSF) and the U.S. Geological Survey (USGS) under NEHRP. In recent fiscal years, neither agency has explicitly summarized its expenditures in this particular task area, but current investments can be estimated from reported agency budgets.

- Significant support by NSF for research on the physics of earthquake processes is channeled through the International Research Institutions for Seismology (IRIS) (total budget of $12.4 million in FY2010), the Southern California Earthquake Center (SCEC) ($3.0 million), and EarthScope ($25.0 million), as well as through NSF's Division of Earth Sciences (EAR) core program in geophysics. At least $15 million of these FY2010 funds supported basic research on earthquake physics.
- The USGS Earthquake Hazards Program expended a total of $13 million on earthquake physics research in FY2010; this amount included $10.6 million for its internal program and $2.4 million for its external programs.

Therefore, FY2010 NEHRP expenditures in support of Task 1 totaled more than $27 million/year, when summed over NSF and USGS. Many of the tasks outlined in this report require a better understanding of earthquake physics. Basic research in this area is proceeding vigorously, as described in Chapter 3, and current levels on investment should be maintained for at least the next 5 years, which implies a minimum 5-year budget of ~$135 million. Following this initial investment, we estimate average annual expenditure of ~$30 million/year.

Task 2—Advanced National Seismic System

- The capitalization cost for the full ANSS is estimated at $175 million. Prior to ARRA and through FY2009, USGS will have invested about

$26 million, and after the ARRA expenditure of $19 million—for a total of $45 million—the system will be about 25 percent complete at the end of 2011.[1]

- Current ANSS operations cost $24 million/year, and operational costs are estimated as $50 million/year when ANSS is fully implemented. The current USGS long-term budget request for ANSS is $50 million/year. Because operational costs will increase as the network is developed, it will become increasingly difficult to allocate sufficient capitalization funds for the network to be completed by the target date of 2018 unless there is a substantially increased funding allocation by Congress.

- These cost estimates include continued support at existing levels for the Global Seismic Network, an important subsystem of ANSS, currently funded under NEHRP at $9.8 million/year ($5.8 million/year by USGS and $4 million/year by NSF).

- These costs also do not include geodetic monitoring, primarily by GPS and strainmeter networks, which is complementary to seismic monitoring. In FY2009, USGS spent $2.35 million on geodetic data collection, which included network operations. NSF supports geodetic data collection, including network operations, primarily through UNAVCO, which received $3.7 million for this purpose in FY2009. Additional support for GPS geodesy comes from NASA.

- It is likely that the ANSS Steering Committee will soon recommend that geodetic networks be incorporated into ANSS, and this will obviously increase the scope and cost of ANSS.

Task 3—Earthquake Early Warning

The implementation of effective earthquake early warning (EEW) systems will require the full implementation of ANSS, and the budget analysis presented here assumes a full implementation.

- Current development activities are limited to the USGS EEW demonstration project in California, which expended $0.5 million in FY2010. The President's request to Congress for EEW is $1 million in FY2011.

- The costs of a 3-year implementation plan for EEW in California have been estimated by the California Integrated Seismic Network to be $53.4 million. This includes $32.4 million for equipment upgrades, new equipment, and software development, and $21 million for product development, development of public and professional best practices, and management. Operational costs for the California EEW system are estimated to be $8 million/year.

---

[1] See earthquake.usgs.gov/monitoring/anss/documents.php.

• Implementation of an EEW system for Cascadia can leverage on existing and planned elements of ANSS and the tsunami warning system. Based on a 3-year development timeline, a rough estimate of the marginal cost is $25 million, about half that of the California system. Operational costs, similarly scaled, would be ~$4 million/year.

• The total 5-year costs for EEW systems in California and Cascadia are estimated to be $103 million.

Task 4—National Seismic Hazard Model

• A table listing annualized costs for years 1-5 ($42.3 million/year), 6-10 ($43.2 million/year), and 11-20 ($37.4 million/year) is presented as Table E.1 in Appendix E.

• The costs of seismic hazard mapping are reported here, but it should be noted that this component contributes substantially to many other tasks, particularly Tasks 13 and 14.

• The total 5-year costs for local and national mapping of seismic hazard are estimated to be approximately $250 million.

Task 5—Operational Earthquake Forecasting

• USGS and NSF are currently supporting the Working Group on California Earthquake Probabilities (WGCEP) to develop the Uniform California Rupture Earthquake Forecast 3 (UCERF3), which will include a short-term forecasting capability, at a rate of approximately $2 million/year. WGCEP is also receiving $0.8 million/year from the California Earthquake Authority. A comparable level of expenditure would be needed to develop earthquake forecasting models in California and other seismically active regions of the United States.

• The President's FY2011 budget request to Congress allocates $3 million for the production of earthquake information at the National Earthquake Information Center in Golden, CO. It also requests $0.5 million to enhance the USGS program in operational earthquake forecasting.

• The costs of prospective testing of operational earthquake forecasts by Collaboratory for the Study of Earthquake Predictability (CSEP) are estimated to be $0.5 million/year.

• The total 5-year costs for operational earthquake forecasting are estimated to be approximately $25 million.

Task 6—Earthquake Scenarios

• The overall cost of producing an earthquake scenario and exercise for an individual community provides the benchmark for the national scale budget estimates presented here. The Fedral Emergency Management Agency (FEMA) Authorized Equipment List (AEL) study identified 43 high-risk communities in the United States with AEL greater than $10 million (FEMA, 2008; see Table 3.2), comprising almost 30 percent of the U.S. population base.

• Experience from conducting the pilot earthquake scenarios indicates that the level of effort is, in part, dictated by the size of the community. Small communities with populations less than 500,000 people, such as the Evansville, IN, example described in Chapter 2, have been able to map the local geology and site conditions, develop GIS databases for Urban Seismic Hazard Maps, improve local building and critical infrastructure inventories, and run loss estimation models for scenario events for ~$0.5 million over a period of 5 years under the USGS Urban Hazard Mapping Program.

• There are 18 high-risk communities with populations of 500,000 or less. Cities with populations greater than 1 million would require proportionally more time and resources. The Saint Louis Urban Hazard Mapping Project, for example, has a mapping program for 29 quadrangles over 10 years. Costs associated with this effort are estimated to be ~$2 million.

• Note that estimates for the Evansville, IN, and Saint Louis, MO, examples do not include costs for conducting community-wide earthquake exercises.

• Larger efforts, such as the 2008 southern California ShakeOut exercise discussed in Chapter 1, involved the NEHRP agencies as well as widespread participation by local scientific, community, and media organizations. The initial "start up costs" for the ShakeOut scenario development and exercise totaled ~$6 million (L. Jones and M. Benthien, written communication, 2011).

• Nationally, there are 16 high-risk communities with populations greater than 1 million.

• Consequently, we estimate it would require ~$200 million to develop a uniform series of urban seismic hazard and risk maps and to conduct earthquake exercises for the 43 communities identified in Table 3.2. Funding for the development of comprehensive earthquake risk scenarios and risk assessments in the current (FY2009) NEHRP budget is $1.5 million; we estimate that $10 million/year will be required.

Task 7—Earthquake Risk Assessment and Applications

• At the national level, support for the development of hazards and risk assessment methodologies and support for the basic research that provides the various elements required for the methodology has been a key element of the NEHRP program. At present (FY2009), support for the development of advanced loss estimation and risk assessment tools in the NEHRP budget is $0.5 million.

- Development of the next generation hazard loss estimation tool—although Hazards U.S. (HAZUS) is useful as an inexpensive and easy-to-use loss estimation tool, and is able to yield approximate estimates of hazard losses, greater accuracy is needed for the focused allocation of funding for loss reduction and for policy decisions in general. This program would take advantage of the significant advances in hazard loss estimation achieved by the three existing Earthquake Engineering Research Centers over the past dozen years, to synthesize these advances and develop an expert system for higher-level use. The goal is software that would be accessible to expert teams addressing strategic decisions and more severe disasters.

- We estimate that the funding required for both short-term methodology development and longer-term capability development is $5 million/year.

Task 8—Post-earthquake Social Science Response and Recovery Research

- Development of Standardized Data Protocols, to include 2-4 methodological projects during the initial 2 years to develop standardized research protocols for social science studies of post-disaster response and recovery activities and preparedness practices associated with them. The cost of these projects and resulting workshops are estimated at $1.5 million.

- Establishment of a National Center for Social Science Research on Earthquakes and Other Disasters—the center's primary mission would be to oversee the implementation of standardized research protocols and address, on a continuing basis, related data management issues. The estimated funding for such a center is $2.3 million/year for the initial 5 years; funding during the remaining 15 years of the plan would be based on a performance review after 5 years.

Task 9—Post-earthquake Information Management

- The cost estimates for a post-earthquake information management system (PIMS) are based on a two-phase development approach (PIMS Project Team, 2008).

- The first phase would develop an initial PIMS capability and could be accomplished in 2 years at $1 million/year.

- The second phase could take from 5 to 10 years and would involve development of a more advanced, "full-function" PIMS. Phase 2 will involve about 7 to 9 pilot projects that would have both a development phase and an implementation phase. Operations costs would continue beyond the development period of Phase 2.

•   There would be substantial additional costs incurred whenever the system is activated post-event to harvest, distribute, and archive information. These costs are beyond the focus of this study and would have to be addressed on a case-by-case basis. A more detailed implementation budget is included with assumptions as Table E.2 in Appendix E.

Task 10—Socio-economic Research on Hazard Mitigation and Recovery

The task includes five research program elements that together total $3 million/year:

•   Research program on mitigation and recovery, to include studies on the cost and effectiveness of various resilient strategies and the use of these results to inform and develop prospective indices of resilience; estimated at $1 million/year. This program would also include evaluation of the role of the new business continuity industry as a complement to government assistance, deeper analysis of organizational response to disasters and obstacles to implementation of resilience, as well as policy instruments to overcome these obstacles and to promote best practice. It would also involve analysis of long-run effects of disasters and comprehensive planning frameworks to promote resilience against any such losses. Research should also be extended into new areas such as equity and justice, and ecological resilience.

•   Research program on the long-term impacts of disasters; estimated at $0.5 million/year. This would involve the further development of a framework for analysis, and rigorous testing at sites of major earthquakes and other major disasters. This program would also address key policy issues including such questions as the necessity of re-building in the same locations, migration support, and mandating of mitigation during the recovery and reconstruction processes.

•   Research program on equity and justice in hazard resilience; estimated at $0.5 million/year. Research would focus on the exploration of equity/justice principles, analysis of the implications of their application, and their acceptance by communities and policy-makers. It would be applied to a broad range of disadvantaged groups including racial/ethnic minorities, women, the aged and the very young, the physically challenged, and the poor.

•   Development of a National Clearinghouse for Economic Resilience; estimated at $1 million/year. This clearinghouse would combine research and practice—research to develop resilience metrics and new resilience strategies that would then be transformed into operational activities and tested in pilot programs. Practitioners in the private and public sectors would share their experiences with the broad community through the clearinghouse. See also an expanded role for Task 11.

Task 11—Observatory Network on Community Resilience and Vulnerability

- Costs associated with development of an Observatory Network on Community Resilience and Vulnerability are estimated to total $14.5 million over the next 5 years (see details in Table E.3 in Appendix E), with continuing funding through Year 20 of $2.9 million/year. This estimate, based on the phased implementation outlined in the RAVON workshop report (Peacock et al., 2008), represents the middle of the cost range suggested in that report.
- Although implementing Tasks 8, 9, 10, and 11 should be considered separately, the potential for leveraging resources across these tasks is substantial. Because of its more global nature, Task 11 would serve as the umbrella for considering such leveraging.

Task 12—Physics-based Simulations of Earthquake Damage and Loss

- The annualized cost for years 1-20 of $6 million/year includes three components: earthquake science ($2 million/year), earthquake engineering ($2 million/year), and information technology ($2 million/year). Funding for the basic science and engineering tasks required to support, improve, and "operationalize" end-to-end simulation tools are included in Tasks 1, 13, 14, and 16.
- Funding for the high-performance computing equipment required to enable end-to-end simulations is assumed to be available through federal agencies or through universities and facilities funded by federal agencies.

Task 13—Techniques for Evaluation and Retrofit of Existing Buildings

- A table listing annualized costs for years 1-5 ($22.9 million/year), 6-10 ($34 million/year), and 11-20 ($26 million/year) is presented as Table E.4 in Appendix E, and a more detailed breakdown for each component—including component timing—is presented in Table E.5.
- Program coordination and management costs are 20 percent of the combined research, development, and implementation costs for this task, distributed uniformly over the full 20 years.

- The costs for NEES operations and maintenance, a substantial contributor to this task, are reported under Task 14.
- The costs for seismic hazard analysis, a key contributor to this task, are reported under Task 4.

Task 14—Performance-based Earthquake Engineering for Buildings

- A table listing annualized costs for years 1-5 ($46.7 million/year), 6-10 ($47.7 million/year), and 11-20 ($41.9 million/year) is presented as Table E.6 in Appendix E, and a more detailed breakdown for each component—including component timing—is presented in Table E.7.
- Program coordination and management costs are 20 percent of the combined research, development, and implementation costs for this task, distributed uniformly over the full 20 years.
- The costs of NEES operations and maintenance are reported here, but it should be noted that the NEES component contributes substantially to many other tasks, particularly Tasks 13 and 16.
- The costs associated with deploying and maintaining ANSS and the costs of seismic hazard analysis, which are key contributors to this task, are reported under Tasks 2 and 4, respectively.

Task 15—Guidelines for Earthquake-Resilient Lifelines Systems

- Both the National Institute of Standards and Technology (NIST) (1997) and EERI (2003b) estimated $3 to $5 million annual budgets for the development of guidelines, manuals of practice, and model codes for seismic design and retrofit of buildings, lifelines, bridges, and coastal structures. EERI (2003b) also identified an additional $5 million/year for demonstration projects and $5 million/year for basic lifeline engineering research.
- Based in part on this background information, we estimate that accomplishing the task as outlined in Chapter 3 would require $5 million/year, representing a very substantial increase from the existing funding level of ~$100,000/year.

Task 16—Next Generation Sustainable Materials, Components, and Systems

- A table listing annualized costs for years 1-5 ($8.2 million/year), 6-10 ($13.9 million/year), and 11-20 ($22.4 million/year) is presented as Table E.8 in Appendix E, and a more detailed breakdown for each component—including component timing—is presented in Table E.9.

- The costs for NEES operations and maintenance, a substantial contributor to this task, are reported under Task 14.

Task 17—Knowledge, Tools, and Technology Transfer to Public and Private Practice

- Annual costs include the development of seismic standards and the development of research consolidation documents ($8.4 million/year), for a total of $168 million over 20 years.

Task 18—Earthquake-Resilient Communities and Regional Demonstration Projects

- The resources that would be needed at any particular time would depend on the number of communities selected, the amount of matching funds provided, and the number and nature of demonstration projects. We recommend that the program begin with a few communities, and then expand as capacity improves and community leaders are developed who can provide peer-to-peer mentoring.
- The average unit cost per community would be about $750,000/ year, varying depending on the size and complexity of each community and the nature of selected demonstration projects. We propose initial funding for the first 2 years at $4 million/year, increasing to $69 million/year per year when the program includes a full complement of 60 communities. Additional cost breakdown information is presented in Table E.10 in Appendix E.

5

Conclusions—
Achieving Earthquake Resilience

The advent of NEHRP in 1977, together with its subsequent reautho-
rizations, added substantial resources for research in seismology,
earthquake engineering, and social sciences with the goal of increasing knowledge for understanding the causes of earthquakes and reducing
their impacts. In addition, the program improved coordination among
federal government agencies with responsibilities in those areas and
promoted integration of research and applications. Moreover, although
NEHRP covers only four federal agencies, the program provides a focus
for earthquake-related activities of many other federal, state, regional,
and local government agencies, and—to some extent—the private sector.

Efforts to understand the causes of earthquakes and to counter their
effects certainly did not begin with NEHRP. In the United States, the
landmark study of the 1906 San Francisco earthquake (Lawson, 1908)
furthered the elastic rebound hypothesis, whereby accumulated strain
energy is released suddenly by fault slip, and demonstrated the vulnerability of structure built on soft sediments. Advances in other countries,
especially Japan, also contributed new knowledge. Most importantly,
developments of plate tectonics concepts in the mid-1960s established an
overall framework for understanding the occurrence of earthquakes (and
volcanoes) worldwide.

Nevertheless, NEHRP stimulated substantial earthquake research
in the United States and, most significantly, integrated the efforts of the
various earthquake-related disciplines and organizations toward the goal
of reducing earthquake losses. The degree of success of these endeavors
is reflected in the impressive list of accomplishments summarized in the

introduction to this report. **In view of the important stimulus to earthquake mitigation activities provided by NEHRP and its substantial record of achievements, the committee endorses the 2008 NEHRP Strategic Plan and identifies 18 specific task elements required to implement that plan and materially improve national earthquake resilience.**

Defining Earthquake Resilience

A critical requirement for achieving national earthquake resilience is, of course, an understanding of what constitutes earthquake resilience. In this report, we have interpreted resilience broadly so that it incorporates engineering/science (physical), social/economic (behavioral), and institutional (governing) dimensions. Resilience is also interpreted to encompass both pre- and post-disaster actions that, in combination, will enhance the robustness and the capabilities of all earthquake-vulnerable regions of our nation to function well following likely, significant earthquakes. The committee is also cognizant that it is cost-prohibitive to achieve a completely seismically resistant nation. Instead, we see our mission as helping set performance targets for improving the nation's seismic resilience over the next 20 years and, in turn, developing a more detailed roadmap and program priorities for NEHRP. With these considerations in mind, the committee recommends that NEHRP adopt the following working definition for "national earthquake resilience":

A disaster-resilient nation is one in which its communities, through mitigation and pre-disaster preparation, develop the adaptive capacity to maintain important community functions and recover quickly when major disasters occur.

No standard metric exists for measuring disaster resilience, and it is clear that standardized methods would be helpful for gauging improvements in resilience as a result of disaster risk reduction planning and mitigation. However, because the concept of resilience is specific to the context of the specific community and its goals, it can be expected that no single measure will be able to capture it sufficiently. No one resilience indicator can suit all purposes, and different measurement approaches may be appropriate in different contexts for assessing current levels of disaster resilience and incremental progress in developing resilience.

Elements and Costs of a Resilience Roadmap

To provide a sound basis for future activities, the NEHRP agencies—under the leadership of the National Institute of Standards and

Technology (NIST) as lead agency—developed a Strategic Plan (Appendix A). The plan, with three major goals and 14 objectives, constitutes a comprehensive, integrated approach to reducing earthquake losses. **The committee endorses the elements of the strategic plan—the goals and objectives—and embraces the integrated, comprehensive, and collaborative approach among the NEHRP agencies reflected in the plan.** The committee set out to build on the Strategic Plan by specifying focused activities that would further implementation of the plan. In the end, 18 tasks were selected, ranging from basic research to community-oriented applications, which, in our view, comprise a "roadmap" for furthering NEHRP goals and implementing the Strategic Plan. The committee recommends that these tasks be undertaken.

In estimating costs to implement the roadmap, the committee recognizes that there is a high degree of variability among the 18 tasks—some (e.g., deployment of the Advanced National Seismic System Network [ANSS], the Network for Earthquake Engineering Simulation [NEES] earthquake engineering simulation laboratories) are under way or are in the process of being implemented, whereas others are only at the conceptual stage. Costing each task required a thorough analysis to determine scope, implementation steps, and linkages or overlaps with other tasks. For some of the tasks, the necessary analysis had already been completed in workshops or other venues, and realistic cost estimates were available as input to the committee. For other tasks, the committee had nothing more to go on than its own expert opinion, in which case implementing the task may require some degree of additional detailed analysis. In summary, the annualized cost for the first 5 years of the roadmap for national earthquake resilience is $306.5 million/year (2009$), made up of the following tasks:

1. **Physics of Earthquake Processes.** Conduct additional research to advance the understanding of earthquake phenomena and generation processes and to improve the predictive capabilities of earthquake science; 5-year annualized cost of $27 million/year, for a total 20-year cost of $585 million.

2. **Advanced National Seismic System.** Complete deployment of the remaining 75 percent of the Advanced National Seismic System; 5-year annualized cost of $66.8 million/year, for a total 20-year cost of $1.3 billion. On-going operations and maintenance costs after the initial 20-year period of $50 million/year.

3. **Earthquake Early Warning.** Evaluation, testing, and deployment of earthquake early warning systems; 5-year annualized cost of $20.6 million/year, for a total 20-year cost of $283 million.

4. **National Seismic Hazard Model.** Complete the national coverage

of seismic hazard maps and create urban seismic hazard maps and seismic risk maps for at-risk communities; 5-year annualized cost of $50.1 million/year, for a total 20-year cost of $946.5 million.

5. **Operational Earthquake Forecasting.** Develop and implement operational earthquake forecasting, in coordination with state and local agencies; 5-year annualized cost of $5 million/year, for a total 20-year cost of $85 million. On-going operations and maintenance costs after the initial 20-year period are unknown.

6. **Earthquake Scenarios.** Develop scenarios that integrate earth science, engineering, and social science information and conduct exercises so that communities can visualize earthquake and tsunami impacts and mitigate their potential effects; 5-year annualized cost of $10 million/year, for a total 20-year cost of $200 million.

7. **Earthquake Risk Assessments and Applications.** Integrate science, engineering, and social science information in an advanced GIS-based loss estimation platform to improve earthquake risk assessments and loss estimations; 5-year annualized cost of $5 million/year, for a total 20-year cost of $100 million.

8. **Post-earthquake Social Science Response and Recovery Research.** Document and model the mix of expected and improvised emergency response and recovery activities and outcomes to improve pre-disaster mitigation and preparedness practices at household, organizational, community, and regional levels; 5-year annualized cost of $2.3 million/year, reviewed after the initial 5-years.

9. **Post-earthquake Information Management.** Capture, distill, and disseminate information about the geological, structural, institutional, and socioeconomic impacts of specific earthquakes, as well as post-disaster response, and create and maintain a repository for post-earthquake reconnaissance data; 5-year annualized cost of $1 million/year, for a total 20-year cost of $14.6 million. On-going operations and maintenance costs after the initial 20-year period are unknown, but are likely to be small.

10. **Socioeconomic Research on Hazard Mitigation and Recovery.** Support basic and applied research in the social sciences to examine individual and organizational motivations to promote resilience, the feasibility and cost of resilience actions, and the removal of barriers to successful implementation; 5-year annualized cost of $3 million/year, for a total 20-year cost of $60 million.

11. **Observatory Network on Community Resilience and Vulnerability.** Establish an observatory network to measure, monitor, and model the disaster vulnerability and resilience of communities, with a focus on resilience and vulnerability; risk assessment, perception, and management strategies; mitigation activities; and reconstruction and recovery; 5-year

annualized cost of $2.9 million/year, for a total 20-year cost of $57.3 million. On-going operations and maintenance costs after the initial 20-year period are unknown.

12. **Physics-based Simulations of Earthquake Damage and Loss.** Integrate knowledge gained in Tasks 1, 13, 14, and 16 to enable robust, fully coupled simulations of fault rupture, seismic wave propagation through bedrock, and soil-structure response, to compute reliable estimates of financial loss, business interruption, and casualties; 5-year annualized cost of $6 million/year, for a total 20-year cost of $120 million.

13. **Techniques for Evaluation and Retrofit of Existing Buildings.** Develop analytical methods that predict the response of existing buildings with known levels of reliability based on integrated laboratory research and numerical simulations, and improve consensus standards for seismic evaluation and rehabilitation; 5-year annualized cost of $22.9 million/year, for a total 20-year cost of $543.6 million.

14. **Performance-based Earthquake Engineering for Buildings.** Advance performance-based earthquake engineering knowledge and develop implementation tools to improve design practice, inform decision-makers, and revise codes and standards for buildings, lifelines, and geo-structures; 5-year annualized cost of $46.7 million/year, for a total 20-year cost of $891.5 million.

15. **Guidelines for Earthquake-Resilient Lifeline Systems.** Conduct lifelines-focused collaborative research to better characterize infrastructure network vulnerability and resilience as the basis for the systematic review and updating of existing lifelines standards and guidelines, with targeted pilot programs and demonstration projects; 5-year annualized cost of $5 million/year, for a total 20-year cost of $100 million.

16. **Next Generation Sustainable Materials, Components, and Systems.** Develop and deploy new high-performance materials, components, and framing systems that are green and/or adaptive; the 5-year annualized cost of $8.2 million/year, for a total 20-year cost of $334.4 million.

17. **Knowledge, Tools, and Technology Transfer to/from the Private Sector.** Initiate a program to encourage and coordinate technology transfer across the NEHRP domain to ensure the deployment of state-of-the-art mitigation techniques across the nation, particularly in regions of moderate seismic hazard; 5-year annualized cost of $8.4 million/year, for a total 20-year cost of $168 million.

18. **Earthquake-Resilient Community and Regional Demonstration Projects.** Support and guide community-based earthquake resiliency pilot projects to apply NEHRP-generated and other knowledge to improve awareness, reduce risk, and improve emergency preparedness and recovery capacity; 5-year annualized cost of $15.6 million/year, for a total 20-year cost of $1 billion.

Timing of Roadmap Components

The committee recommends that all the tasks identified here be initiated immediately, contingent on the availability of funds, and suggests that such an approach would represent an appropriate balance between practical activities to enhance national earthquake resilience and the research that is needed to provide a sound basis for such activities. The committee also notes that the two "observatory" elements of the roadmap, Task 2 and Task 11, will provide fundamental information to be used by numerous other tasks.

However, at a lower component level within individual tasks, there are some elements that should be implemented and/or initiated immediately whereas others will have to await the results of earlier activities. The need for sequencing individual task components is most clearly expressed in the detailed breakdowns for Tasks 13, 14, and 16, as described in Tables E.5, E.7, and E.9 respectively. For example, the component to develop reliable tools for collapse computations within Task 13 includes scoping studies, a workshop, and development of a work-plan in year 3 that would be followed by experimentation using NEES facilities on critical components of framing systems in years 4-7, experimentation using NEES facilities and E-Defense on multiple framing systems to collapse in years 6-10, and concurrent development of improved hysteretic models of structural components through failure in years 4-20, understanding of the triggers for collapse of framing systems in years 6-10, improved system-level collapse computations and FE codes in years 6-15, validation of improved computational procedures using NEES facilities and E-Defense in years 11-20, as well as 5-yearly syntheses of results and preparation of technical briefs.

Earthquake Resilience and Agency Coordination

It is important to recognize that the four NEHRP agencies, although comprising a critical core group for building earthquake knowledge, constitutes only part of the national research and application enterprise. For example, the National Science Foundation (NSF) part of NEHRP includes only earthquake engineering and social sciences, viewed by NSF as "directed" research, whereas highly relevant earthquake knowledge also comes from "non-directed" research programs in NSF. In the applications area, virtually every agency that builds or operates facilities contributes to the goals of NEHRP by adopting practices or codes to reduce earthquake impacts. These agencies include the U.S. Army Corps of Engineers and the Departments of Transportation, Energy, and Housing and Urban Development. Beyond the role of the federal agencies, government agencies at all levels similarly play a critical role in application of earthquake knowledge, as does the private sector, especially in the area of building design. Altogether, the contributors to reducing earthquake losses constitute a complex enterprise that goes far beyond the scope of NEHRP. But NEHRP provides an important focus for this far-flung endeavor. The committee considers that an analysis to determine whether coordination among all organizations that contribute to NEHRP could be improved would be useful and timely.

Implementing NEHRP Knowledge

The United States had not experienced a great earthquake since 1964, when Alaska was struck by a magnitude-9.2 event. The damage in Alaska was relatively light because of the sparse population. The 1906 San Francisco earthquake was the most recent truly devastating U.S. shock, as recent destructive earthquakes have been only moderate in size. Consequently, a sense has developed that the country can cope effectively with the earthquake threat and is, in fact, "resilient." However, coping with moderate events may not be a true indicator of preparedness for a great one, as demonstrated by Hurricane Katrina. The central United States last experienced a devastating sequence of great earthquakes in 1811-1812 in the Mississippi Valley area centered on New Madrid, MO. The East Coast was shocked in 1886 by an earthquake near magnitude-7 at Charleston, SC. These events are now far from the consciousness of the public, and little has been done to prepare for similar events in these regions in the future. The committee believes that efforts should be expanded to anticipate the effects and disruptions that could be caused by a great U.S. earthquake, especially an event in the central or eastern United States where little preparation has been undertaken.

Most critical decisions that reduce earthquake vulnerability and manage earthquake risk are made in the private sector by individuals and companies. The information provided by NEHRP, if made available in an understandable format, and accompanied by diffusion processes, can greatly assist citizens in their decision-making. For example, maps of active faults, unstable ground, and historic seismicity can influence where people choose to live, and maps of relative ground shaking can guide building design.

NEHRP will have accomplished its fundamental purpose—an earthquake-resilient nation—when those responsible for earthquake risk and for managing the consequences of earthquake events use the knowledge and services created by NEHRP and other related endeavors to make our communities more earthquake resilient. Resiliency requires awareness of earthquake risk, knowing what to do in response to that risk, and doing it. But providing information is not enough to achieve resilience—the diffusion of NEHRP knowledge and implementation of that knowledge are necessary corollaries. Successfully diffusing NEHRP knowledge into communities and among the earthquake professionals, state and local government officials, building owners, lifeline operators, and others who have the responsibility for how buildings, systems, and institutions respond to and recover from earthquakes, will require a dedicated and strategic effort. This diffusion role reflects the limited authority that resides with federal agencies in addressing the earthquake threat. Local and state governments have responsibility for public safety and welfare, including powers to regulate land use to avoid hazards, enforce building codes, provide warnings to threatened communities, and respond to an event. The goals and objectives of NEHRP are aimed at supporting and facilitating measures to improve resilience through private owners and businesses, and supporting local and state agencies in carrying out their duties. Although implementing NEHRP knowledge must move ahead expeditiously, it is also essential that the frontiers of knowledge be advanced in concert, requiring that improving understanding of the earthquake threat, reducing risk, and developing the processes to motivate implementation actions, should all be continuing endeavors.

References

AASHTO (American Association of State Highway and Transportation Officials), 2009. Guide Specifications for the LRFD Seismic Bridge Design, 1st Edition. Washington, DC: AASHTO.

ACEHR (Advisory Committee on Earthquake Hazards Reduction), 2009. Letter to the NIST Deputy Director on the Reauthorization of the NEHRP program. May 4. Available at www.nehrp.gov/pdf/may_2009_letter2.pdf (accessed April 30, 2010).

ACI (American Concrete Institute), 2008. Building Code Requirements for Structural Concrete and Commentary. ACI-318-08. Farmington Hills, MI: ACI.

ALA (American Lifelines Alliance), 2003. Existing Guidelines Matrix. Available at www.americanlifelinesalliance.org/ExistingGuidelines.htm (accessed April 30, 2010).

Alesch, D.J., L.A. Arendt, and J.N. Holly, 2009. Managing for Long-Term Community Recovery in the Aftermath of Disaster. Fairfax, VA: Public Entity Risk Institute.

Algermissen, S.T., 1969. Seismic risk studies in the United States. Paper presented at the Fourth World Conference on Earthquake Engineering, Santiago, Chile, January 13–18.

Allen, R.M., and H. Kanamori, 2003. The potential for earthquake early warning in Southern California. Science 300(5620): 786–789.

Allen, R.M., P. Gasparini, O. Kamigaichi, and M. Böse, 2009. The status of earthquake early warning around the world: an introductory overview. Seismological Research Letters 80: 682–693. DOI:10.1785/gssrl.80.5.682.

ASCE (American Society of Civil Engineers), 2003. Seismic Evaluation of Existing Buildings. ASCE 31-03. Eds. D.B. Horn and C.D. Poland. Proceedings of the 2004 Structures Congress, May 22–26, 2004, Nashville, TN.

ASCE (American Society of Civil Engineers), 2005. Minimum Design Loads for Buildings and Other Structures. SEI/ASCE Standard 7-05. Reston, VA.

ASCE (American Society of Civil Engineers), 2007. Seismic Rehabilitation of Existing Buildings. ASCE/SEI 41-06. Washington, DC.

ASCE (American Society of Civil Engineers), 2009. Report Card for America's Infrastructure: Report Card 2009 Grades. Available at apps.asce.org/reportcard/2009/grades.cfm (accessed April 30, 2010).

ATC (Applied Technology Council), 1991. Seismic Vulnerability and Impact of Disruption of Lifelines in the Conterminous United States. ATC-25. Redwood City, CA.

ATC (Applied Technology Council), 1994. Seismic Evaluation and Retrofit of Concrete Buildings. ATC-40. Redwood City, CA.

ATC (Applied Technology Council), 2007. Prioritized Research for Reducing the Seismic Hazards of Existing Buildings. ATC-73. Redwood City, CA.

ATC (Applied Technology Council), 2009a. Existing Buildings Program Action Plan 2009-2019. ATC-71. Draft in progress. Redwood City, CA.

ATC (Applied Technology Council), 2009b. 50% Draft Guidelines for the Seismic Performance Assessment of Buildings. ATC-58 Report. Draft in progress. Available at www.atcouncil.org (accessed April 30, 2010).

ATC (Applied Technology Council), 2010. Here Today—Here Tomorrow: The Road to Earthquake Resilience in San Francisco. A Community Action Plan for Seismic Safety. ATC-52-2. Redwood City, CA.

Ballantyne, D., 2007. Seattle Fault Earthquake Scenario. Presented at the New Madrid Earthquake Scenario Workshop, St. Louis, MO, April 20. Available at www.eeri.org/site/images/projects/newmadrid/6-nmes-wshop-seattle-scen-balantyne.pdf (accessed April 30, 2010).

Berke, P.R., and T.J. Campanella, 2006. Planning for Postdisaster Resiliency. The ANNALS 604: 192–207.

Berke, P.R., J. Kartez, and D. Wenger, 1993. Recovery after disaster: Achieving sustainable development, mitigation and equity. Disasters 17: 93–109.

Blumstein, C., and S. Wiel, 1999. Public-interest research and development in the electric and gas utility industries. Utilities Policy 7: 191–199

Boettke, P., E. Chamlee-Wright, P. Gordon, S. Ikeda, P. Leson, II, and R. Sobel, 2007. Political, Economic and Social Aspects of Katrina. Southern Economic Journal 74: 363–376.

Borque, L., 2001. TriNet Policy Studies and Planning Activities in Real-Time Earthquake Early Warning: Task 1 Report, Survey of Potential Early Warning System Users. Los Angeles, CA: UCLA.

Bruneau, M., S.E. Chang, R.T. Eguchi, G.C. Lee, T.D. O'Rourke, A.M. Reinhorn, M. Shinozuka, K. Tierney, W.A. Wallace, and D. von Winterfeldt, 2003. A framework to quantitatively assess and enhance the seismic resilience of communities. Earthquake Spectra 19: 733–752.

Burby, R.J., ed., 1998. Cooperating with Nature: Confronting Natural Hazards with Land-Use Planning for Sustainable Communities. Washington, DC: Joseph Henry Press.

Burns, W. J., and P. Slovic, 2007. The diffusion of fear: Modeling community response to a terrorist strike. JDMS: The Journal of Defense Modeling and Simulation: Applications, Methodology, Technology 4: 298–317.

Burns, W.J., R.J. Hofmeister, and Y. Wang, 2008. Geologic hazards, earthquake and landslide hazard maps, and future earthquake damage estimates for six counties in the Mid/Southern Willamette Valley including Yamhill, Marion, Polk, Benton, Linn, and Lane Counties, and the City of Albany, Oregon, Oregon Department of Geology and Mineral Industries, IMS-24, 121 pp., scale 1:422,400.

CARRI (Community and Regional Resilience Institute), 2010. Toward a Common Framework for Community Resilience. Oak Ridge, TN: ORNL.

CGS (California Geological Survey), 1982. Planning Scenario for a Magnitude 8.3 Earthquake on the San Andreas Fault in the San Francisco Bay Area, California. SP061. Sacramento, CA: CGS.

CGS (California Geological Survey), 1987. Planning Scenario for a Magnitude 7.5 Earthquake on the Hayward Fault in the San Francisco Bay Area, California. SP078. Sacramento, CA: CGS.

CGS (California Geological Survey), 1988. Planning Scenario for a Major Earthquake on the Newport-Englewood Fault zone (Los Angeles and Orange Counties, California). SP099. Sacramento, CA: CGS.

CGS (California Geological Survey), 1993. Planning Scenario for a Major Earthquake on the San Jacinto fault in the San Bernardino area. SP100. Sacramento, CA: CGS.

CGS (California Geological Survey), 1995. Planning Scenario in Humboldt and Del Norte Counties for a Great Earthquake on the Cascadia Subduction Zone. SP115. Sacramento, CA: CGS.

Chang, S.E., 2009. Conceptual Framework of Resilience for Physical, Financial, Human, and Natural Capital, School of Community and Regional Planning, University of British Columbia, Burnaby, BC.

Chang, S.E., 2010. Urban disaster recovery: a measurement framework with application to the 1995 Kobe earthquake. *Disasters* 34(2): 303–327.

Chang, S.E. and M. Shinozuka, 2004. Measuring improvements in the disaster resilience of communities. *Earthquake Spectra* 20: 739–755.

Chen, R., D. Branum, and C. Wills, 2009, HAZUS loss estimates for California scenario earthquakes, California Geological Survey.

Comfort, L., 1999. Shared Risk: Complex Seismic Response. New York: Pergamon.

Cox, A., F. Prager and A. Rose. 2011. Transportation security and the role of resilience: A foundation for operational metrics. *Transport Policy* 18: 307–317.

CREW (The Cascadia Region Earthquake Workgroup), 2005. Cascadia Subduction Zone Earthquakes: A Magnitude 9.0 Earthquake Scenario. Publication 0-05-05. Salem, OR: Oregon Department of Geology and Mineral Industries.

Crone, A.J. and R.L. Wheeler, 2000. Data for Quaternary Faults, Liquefaction Features, and Possible Tectonic Features in the Central and Eastern United States, East of the Rocky Mountain Front. U.S. Geological Survey Open-File Report 00-0260. Denver, CO: USGS.

Crowell, B.W., Y. Bock, and M.B. Squibb, 2009. Demonstration of earthquake early warning using total displacement waveforms from real-time GPS networks. *Seismological Research Letters* 80(5): 772–782.

Cutter, S.L., (ed.), 2001. American Hazardscapes: The Regionalization of Hazards and Disasters. Washington, DC: Joseph Henry Press.

Cutter, S.L., B.J. Boruff, and W.L. Shirley, 2003. Social vulnerability to environmental hazards. *Social Science Quarterly* 84: 242–261.

Cutter, S.L., L. Barnes, M. Berry, C. Burton, E. Evans, E. Tate, and J. Webb, 2008a. Community and regional resilience: Perspectives from hazards, disasters, and emergency management. CARRI Research Report 1. Oak Ridge, TN: Oak Ridge National Lab.

Cutter, S.L., L. Barnes, M. Berry, C. Burton, E. Evans, E. Tate, and J. Webb, 2008b. A place-based model for understanding community resilience to natural disasters. *Global Environmental Change* 18: 598–606.

Cutter, S.L., C. Burton, and C. Emrich, 2010. Disaster resilience indicators for benchmarking baseline conditions. *Journal of Homeland Security and Emergency Management* 7: Article 51.

DHS (U.S. Department of Homeland Security), 2006. National Infrastructure Protection Program. Available at www.fas.org/irp/agency/dhs/nipp.pdf (accessed April 30, 2010).

DHS (U.S. Department of Homeland Security), 2009. National Infrastructure Protection Plan: Partnering to Enhance Protection and Resiliency. Washington, DC: U.S. Department of Homeland Security. Available at www.dhs.gov/xlibrary/assets/NIPP_Plan.pdf (accessed August 5, 2010).

Drabek, T.E., 2010. *The Human Side of Disaster.* Boca Raton, FL: CRC Press, Taylor and Francis Group.

EarthScope, 2007. EarthScope Facility Operation and Maintenance: October 1, 2008–September 30, 2018. Proposal to the National Science Foundation. Volume I. Available at www.earthscope.org/es_doc/oandm/OandM_Volume_I.pdf (accessed October 1, 2010).

EERI (Earthquake Engineering Research Institute), 1996. Scenario for a 7.0 Magnitude Earthquake on the Hayward Fault. Oakland, CA: EERI.

EERI (Earthquake Engineering Research Institute), 2003a. Collection & Management of Earthquake Data: Defining Issues for an Action Plan. Oakland, CA: EERI.

EERI (Earthquake Engineering Research Institute), 2003b. Securing Society against Catastrophic Earthquake Losses: A Research and Outreach Plan in Earthquake Engineering. Oakland, CA: EERI.

EERI (Earthquake Engineering Research Institute), 2005. Scenario for a Magnitude 6.7 Earthquake on the Seattle Fault. Oakland, CA: EERI and the Washington Military Department Emergency Management Division.

EERI (Earthquake Engineering Research Institute), 2008. Contributions of Earthquake Engineering to Protecting Communities and Critical Infrastructure from Multihazards. Oakland, CA: EERI.

Ehrlich, I., and G. Becker, 1972. Market insurance, self insurance and self protection. *Journal of Political Economy* 80: 623–648.

Elnashai, A.S., T. Jefferson, F. Fiedrich, L.J. Cleveland, and T. Gress, 2009. Impact of New Madrid Seismic Zone Earthquakes on Central USA. Mid-America Earthquake Center Report, No. 09-03, Vol. 1.

Emmer, R., L. Swann, M. Schneider, S. Sempier, and T. Sempier, 2008. Coastal Resiliency Index: A Community Self-Assessment. MASGP-08-014. Washington, DC: National Oceanic and Atmospheric Administration.

Feinstein, D., 2001. S. 424: A bill to provide incentives to encourage private sector efforts to reduce earthquake losses, to establish a natural disaster mitigation program, and for other purposes. Presented to the Committee on Finance. In Congressional Record—Senate, March 1, 2001, 147(26), Washington, DC: U.S. Government Printing Office, pp. S1754-S1760.

FEMA (Federal Emergency Management Agency), 1985a. An Action Plan for Reducing Earthquake Hazards of Existing Buildings. FEMA 90. Washington, DC.

FEMA (Federal Emergency Management Agency), 1985b. Proceedings of the Workshop on Reducing Seismic Hazards of Existing Buildings. FEMA 91. Washington, DC.

FEMA (Federal Emergency Management Agency), 1995. Plan for Developing and Adopting Seismic Design Guidelines and Standards for Lifelines. FEMA Report 271. Washington, DC.

FEMA (Federal Emergency Management Agency), 1997a. NEHRP Guidelines for the Seismic Rehabilitation of Buildings. FEMA Report 273. Washington, DC.

FEMA (Federal Emergency Management Agency), 1997b. Report on Costs and Benefits of Natural Hazards Mitigation. FEMA Report 294. Washington, DC.

FEMA (Federal Emergency Management Agency), 2001. HAZUS 99 Estimated Annualized Earthquake Losses for the United States. FEMA Report 366. Washington, DC.

FEMA (Federal Emergency Management Agency), 2005. Improvement of Nonlinear Static Seismic Analysis Procedures. FEMA 440. Washington, DC.

FEMA (Federal Emergency Management Agency), 2006. Techniques for the Seismic Rehabilitation of Existing Buildings. FEMA 547. Washington, DC.

FEMA (Federal Emergency Management Agency), 2008. HAZUS-MH Estimated Annualized Earthquake Losses for the United States. FEMA 366. Washington, DC.

FEMA (Federal Emergency Management Agency), 2009a. Quantification of Building Seismic Performance Factors. FEMA Report P-695. Washington, DC.

FEMA (Federal Emergency Management Agency), 2009b. NEHRP Recommended Provisions for New Buildings and Other Structures. FEMA Report P-750. Washington, DC.

Field, E.H., H.A. Seligson, N. Gupta, V. Gupta, T.H. Jordan, and K. Campbell, 2005. Loss estimates for a Puente Hills blind-thrust earthquake in Los Angeles, California, *Earthquake Spectra* 21: 329–338.

Field, E.H., T.E. Dawson, K.R. Felzer, A.D. Frankel, V. Gupta, T.H. Jordan, T. Parsons, M.D. Petersen, R.S. Stein, R.J. Weldon II, and C.J. Wills, 2007. The Uniform California Earthquake Rupture Forecast, Version 2 (UCERF 2). Prepared in cooperation with the California Geological Survey and the Southern California Earthquake Center. USGS Open File Report 2007-1437.

Flynn, S., 2008. America the resilient: Defying terrorism and mitigating natural resources. *Foreign Affairs* 87: 2.

Foster, K.A., 2007. A Case Study Approach to Understanding Regional Resilience. Working Paper 2007–08. Institute of Urban and Regional Development, University of California, Berkeley.

Frankel, A., C. Mueller, T. Barnhard, D. Perkins, E. Leyendecker, N. Dickman, S. Hanson, and M. Hopper, 1996. National Seismic Hazard Mapping Program, National Seismic-Hazard Maps: Documentation June 1996. Open File Report 96-532. Washington, DC: U.S. Geological Survey.

Frankel, A., M.D. Petersen, C.S. Muller, K.M. Haller, R.L. Wheeler, E.V. Leyendecker, R.L. Wesson, S.C. Harmsen, C.H. Cramer, D.M. Perkins, and K.S. Rukstales, 2002. National Seismic Hazard Mapping Program, Documentation for the 2002 Update of the National Seismic Hazard Maps. Open File Report 2002-420. Washington, DC: U.S. Geologic Survey.

Gerstenberger, M., L.M. Jones, and S. Wiemer, 2007. Short-term aftershock probabilities: case studies in California. *Seismological Research Letters* 78: 66–77. DOI: 10.1785/gssrl.78.1.66.

Giesecke, J., W.J. Burns, A. Barrett, E. Bayrak, A. Rose, and M. Suher, 2010. Assessment of the Regional Economic Impacts of Catastrophic Events: CGE Analysis of Resource Loss and Behavioral Effects of an RDD Attack Scenario. *Risk Analysis*, forthcoming.

Gigerenzer, G., 2004. Fast and frugal heuristics: The tools of bounded rationality. Pp. 62–88 in D. Koehler and N. Harvey (eds.), Blackwell Handbook of Judgment and Decision Making. Oxford, England: Blackwell.

Gladwell, M., 2000. *The Tipping Point: How Little Things Can Make a Big Difference*. New York: Little, Brown and Company.

Godschalk, D.R., 2003. Urban hazard mitigation: Creating resilient cities. *Natural Hazards Review* 4: 136–143.

Godschalk, D.R., A. Rose, E. Mittler, K. Porter, and C.T. West, 2009. Estimating the value of foresight: Aggregate analysis of natural hazard benefits and costs. *Journal of Environmental Planning and Management* 52: 739–756.

Graves, R., S. Callaghan, P. Small, G. Mehta, K. Milner, G. Juve, K. Vahi, E. Field, E. Deelman, D. Okaya, P. Maechling, and T.H. Jordon, 2010. Full waveform physics-based probabilistic seismic hazard calculations for Southern California using the SCEC Cyber Shake platform. The University of Tokyo, Symposium on Long Period Ground Motion and Urban Disaster Mitigation, Japan, March 17–18, 2010.

Grubesic, T.H., T.C. Matisziw, A.T. Murray, and D. Snediker, 2008. Comparative approaches for assessing network connectivity and vulnerability. *International Regional Science Review* 31: 88–112.

Haimes, Y., 2009. On the definition of resilience in system. *Risk Analysis* 29: 498–501.

Hammond, W.C., B.A. Brooks, R. Bürgmann, T. Heaton, M. Jackson, and A.R. Lowry. 2010. The Scientific Value of High-Rate, Low-Latency GPS Data. A White Paper by the Earth-Scope Plate Boundary Observatory Advisory Committee. August. Available at www. unavco.org/research_science/science_highlights/2010/RealTimeGPSWhitePaper2010. pdf (accessed October 4, 2010).

Heal, G., and H. Kunreuther, 2007. Modeling interdependent risks. *Risk Analysis* 27: 621–633.

Heikkla, E., and Y. Wang, 2009. Fujita and Ogawa revisited: an agent-based modeling approach. *Environment and Planning B: Planning and Design* 36: 741–756.

Helz, R.L., 2005. Monitoring Ground Deformation from Space. Fact Sheet 2005-3025. July. Reston, VA: U.S. Geological Survey.

Holmes, W.T., 2002. Background and History of the Seismic Hospital Program in California. Proceedings, Seventh U.S. National Conference on Earthquake Engineering, July 21-25, 2005, Boston, MA. Earthquake Engineering Research Institute, Oakland, CA.

ImageCat, Inc., and ABS Consulting, 2006. Data Standardization Guidelines for Loss Estimation–Populating Inventory Databases for HAZUS®MH MR-1. Prepared for the California Governor's Office of Emergency Services. Available at www.usehazus.com/docs/loss_estimation_guide.pdf (accessed February 9, 2011).

Jackson, T., 2005. Live better by consuming less? Is there a "double dividend" in sustainable consumption? *Journal of Industrial Ecology* 9: 19–36.

Johnson, L.A., 2009. Developing a Management Framework for Local Disaster Recovery: A study of the U.S. disaster recovery management system and the management processes and outcomes of disaster recovery in 3 U.S. cities. Dissertation submitted in partial fulfillment of the Doctoral Degree, School of Informatics, Kyoto University. March.

Jones, L.M., 1991. Short-term Earthquake Hazard Assessment for the San Andreas Fault in southern California. U.S. Geological Survey Open File Report 91–32.

Jones, L.M., R. Bernknopf, D. Cox, J. Goltz, K. Hudnut, D. Mileti, S. Perry, D. Ponti, K. Porter, M. Reichle, H. Seligson, K. Shoaf, J. Treiman, and A. Wein, 2008. The ShakeOut Scenario: U.S. Geological Survey Open-File Report 2008-1150 and California Geological Survey Preliminary Report 25. Available at pubs.usgs.gov/of/2008/1150/ (accessed August 13, 2010).

Jordan, T.H., and L.M. Jones, 2010. Operational earthquake forecasting: some thoughts on why and how. *Seismological Society of America* 81: 571–574.

Jordan, T.H., Y.-T. Chen, P. Gasparini, R. Madariaga, I. Main, W. Marzocchi, G. Papadopoulos, G. Sobolev, K. Yamaoka and J. Zschau, 2009. *Operational Earthquake Forecasting: State of Knowledge and Guidelines for Implementation*. Findings and Recommendations of the International Commission on Earthquake Forecasting for Civil Protection, Dipartimento della Protezione Civile, Rome, Italy, October 2, 2009.

Kajitani, Y., and H. Tatano, 2007. Estimation of lifeline resilience factors based on empirical surveys of Japanese industries. *Earthquake Spectra* 25: 755–776.

Kamigaichi, O., M. Saito, K. Doi, T. Matsumori, S. Tsukada, K. Takeda, T. Shimoyama, K. Nakamura, M. Kiyomoto, and Y. Watanabe, 2009. Earthquake early warning in Japan: Warning the general public and future prospects. *Seismology Research Letters* 80: 717–726.

Kircher, C.A., H.A. Seligson, J. Bouabid, and G.C. Morrow, 2006. When the Big One strikes again—estimated losses due to a repeat of the 1906 San Francisco earthquake. *Earthquake Spectra* 22: S297–S339.

Kreps, G.A., 2001. "Disasters, sociology of." Pp. 3719–3721 in N. Smelser and P. Bates (eds.), International Encyclopedia of the Social and Behavioral Sciences, Vol. 6. Oxford, UK: Elsevier.

Kreps, G.A., and T.E. Drabek, 1996. Disasters are non-routine social problems. *International Journal of Mass Emergencies and Disasters* 14: 129–153.

Kreps, G.A., and S.L. Bosworth, 2006. Organizational adaptation to disaster. Pp. 297–316 in *Handbook of Disaster Research*, H. Rodriguez, E.L. Quarantelli, and R.R. Dynes (eds.), New York: Springer.

Kunreuther, H., and R.R. Roth (eds.), 1998. Paying the Price: The Status and Role of Insurance against Natural Disasters in the United States. Washington, DC: Joseph Henry Press.

Kunreuther, H., R. Ginsberg, L. Miller, P. Sagi, P. Slovic, B. Borkan, and N. Katz, 1978. Disaster Insurance Protection: Public Policy Lessons. New York: Wiley Interscience.

Kunreuther, H., R. Meyer, and C. Van den Bulte, 2004. *Risk Analysis for Extreme Events: Economic Incentives for Reducing Future Losses.* NIST Technical Report GCR 04-871. Washington, DC: National Institute of Standards and Technology.

Kverndokk, S., and A. Rose, 2008. Equity and justice in global warming policy. *International Review of Environmental and Resource Economics* 2(2): 135–176.

Lancieril, M., and A. Zollo, 2008. A Bayesian approach to the real-time estimation of magnitude from the early P and S wave displacement peaks. *Journal of Geophysical Research* 113 (B12302).

Lawson, A.C., 1908. The California Earthquake of April 18, 1906: Report of the State Earthquake Investigation Commission. Publication 87, 2 vols. Washington, DC: Carnegie Institution.

Luco, N., and E. Karaca, 2007. Extending the USGS National Seismic Hazard Maps and ShakeMaps to probabilistic building damage and risk maps. In *Proceedings of the 10th International Conference on Applications of Statistics and Probability in Civil Engineering*, Tokyo, Japan.

McCalpin, J.P., ed., 2009. Paleoseismology (2nd Edition). San Diego, CA: Academic Press.

McDaniels, T., S.E. Chang, D. Cole, J. Mikawoz, and H. Longstaff, 2008. Fostering resilience to extreme events within infrastructure systems: Characterizing decision contexts for mitigation and adaptation. *Global Environmental Change* 18: 310–318.

MCEER (Multidisciplinary Center for Earthquake Engineering Research), 2008. MCEER research: Enabling disaster-resilient communities. *Seismic Waves* (November) 1–2. Available at www.nehrp.gov/pdf/SeismicWavesNov08.pdf (accessed October 4, 2010).

Mendonca, D., 2007. Decision support for improvisation in response to the 2001 World Trade Center attack. *Decisions Support Systems* 43: 952–967.

Miles, S.B., and S.E. Chang, 2006. Modeling community recovery from earthquakes. *Earthquake Spectra* 22: 439–458.

Mileti, D., 1999. Disasters by Design: A Reassessment of Natural Hazards in the United States. Washington, DC: The Joseph Henry Press.

MMC (Multihazard Mitigation Council), 2005. Natural Hazard Mitigation Saves: An Independent Study to Assess the Future Savings from Mitigation Activities. Washington, DC: National Institute of Building Sciences. Available at www.nibs.org/index.php/mmc/projects/nhms/ (accessed June 30, 2010).

Muto, M., S. Krishnan, J.L. Beck, and J. Mitrani-Reiser, 2008. Seismic loss estimation based on end-to-end simulation. Pp. 215–220 in F. Biodini and D.M. Frangopol (eds.), Life-Cycle Civil Engineering. London: CRC Press.

Navrud, S., and R. Ready (eds.) 2002. Valuing Cultural Heritage: Applying Environmental Valuation Techniques to Historic Buildings, Monuments and Artifacts. Northampton, MA: Edward Elgar Publishing.

NEHRP (National Earthquake Hazards Reduction Program), 2007. Program Overview. Available at www.nehrp.gov/pdf/nehrp_acehr_ppt.pdf (accessed August 13, 2010).

NHC (Natural Hazards Center), 2006. Holistic Disaster Recovery: Ideas for Building Local Sustainability after a Natural Disaster. Boulder, Colorado: University of Colorado.

NIBS (National Institute of Building Sciences), 1989. Strategies and Approaches for Implementing a Comprehensive Program to Mitigate the Risk to Lifelines from Earthquakes and Other Natural Hazards. Washington, DC: NIBS. Catalog No. 5047-8.

NIBS (National Institute of Building Sciences), 2007. *American Lifelines Alliance Workshop on Unified Data Collection.* Washington, DC: NIBS. November. Available at www.americanlifelinesalliance.org/pdf/ALAdatawkshprpt.pdf (August 13, 2010).

NIST (National Institute of Standards and Technology), 1996. Proceedings of a Workshop on Developing and Adopting Seismic Design and Construction Standards for Lifelines. NISTIR 5907. Gaithersburg, MD.

NIST (National Institute of Standards and Technology), 1997. Recommendations of the Lifeline Policymakers Workshop. NISTIR 6085. Gaithersburg, MD.

NIST (National Institute of Standards and Technology), 2008. Strategic Plan for the National Hazards Reduction Program: Fiscal Years 2009-2013. Gaithersburg, MD. Available at nehrp.gov/pdf/strategic_plan_2008.pdf (accessed October 4, 2010).

NIST (National Institute of Standards and Technology), 2009. Research Required to Support the Full Implementation of Performance-based Seismic Design. NIST Report GCR 09-917-2. Gaithersburg, MD.

Norris, F.H., S.P. Stevens, B. Pfefferbaum, K.F. Wyche, and R.L. Pfefferbaum, 2008. Community resilience as a metaphor, theory, set of capacities, and strategy for disaster readiness. *American Journal of Community Psychology* 41: 127–150.

NRC (National Research Council), 2001. Review of EarthScope Integrated Science. Washington, DC: National Academy Press.

NRC (National Research Council), 2003. Living on an Active Earth: Perspectives on Earthquake Science. Washington, DC: The National Academies Press.

NRC (National Research Council), 2006a. Facing Hazards and Disasters: Understanding Human Dimensions. Washington, DC: The National Academies Press.

NRC (National Research Council), 2006b. Improved Seismic Monitoring—Improved Decision-Making: Assessing the Value of Reduced Uncertainty. Washington, DC: The National Academies Press.

NRC (National Research Council), 2006c. CLEANER and NSF's Environmental Observatories. Washington, DC: The National Academies Press.

NRC (National Research Council), 2007. Improving Disaster Management: The Role of IT in Mitigation, Preparedness, Response, and Recovery. Washington, DC: The National Academies Press.

NRC (National Research Council), 2009. Sustainable Critical Infrastructure Systems—A Framework for Meeting 21st Century Imperatives. Washington, DC: The National Academies Press.

NRC (National Research Council), 2010. Tsunami Warning and Preparedness: An Assessment of the U.S. Tsunami Program and the Nation's Preparedness Efforts. Washington, DC: The National Academies Press.

O'Rourke, T., 2009. Testimony to the Subcommittee on Technology Innovation. U.S. House of Representatives Committee on Science on the Reauthorization of the National Earthquake Hazards Reduction Program. June 11.

Olsen, K.B., S. Day, J.B. Minster, Y. Cui, A. Chourasia, M. Faerman, R. Moore, P. Maechling, and T. H. Jordan, 2006. Strong Shaking in Los Angeles Expected From Southern San Andreas Earthquake. *Geophysical Research Letters* 33, L07305. DOI:10.1029/2005GL025472.

Olshansky. R., and S. Chang, 2009. Planning for disaster recovery: emerging research needs and challenges. *Journal of Progress in Planning* 200-209.

Olshansky, R., L. Johnson, and K. Topping, 2006. Rebuilding communities following disaster: Lessons from Kobe and Los Angeles. *Built Environment* 32: 354–374.

Peacock, W.G., H. Kunreuther, W.H. Hooke, S.L. Cutter, S.E. Chang, and P.R. Berke, 2008. Toward a Resiliency and Vulnerability Observatory Network: RAVON. College Station, TX: Hazard Reduction and Recovery Center, Texas A&M University. HRRC report 08-02-R.

Perry, S., D. Cox, L. Jones, R. Bernknopf, J. Goltz, K. Hudnut, D. Mileti, D. Ponti, K. Porter, M. Reichle, H. Seligson, K. Shoaf, J. Treiman, and A. Wein, 2008. The ShakeOut Earthquake Scenario—A story that southern Californians are writing. USGS Circular 1324.

Petak, W.J., and A.A. Atkisson, 1982. Natural Hazard Risk Assessment and Public Policy. New York: Springer-Verlag.

Petersen, M.D., A.D. Frankel, S.C. Harmsen, C.S. Mueller, K.M. Haller, R.L. Wheeler, R. L. Wesson, Y. Zeng, O.S. Boyd, D.M. Perkins, N. Luco, E.H. Field, C.J. Wills, and K.S. Rukstales, 2008. Documentation for the 2008 Update of the United States National Seismic Hazard Maps. U.S. Geological Survey Open-File Report 2008–1128. Washington, DC: USGS.

PIMS Project Team, 2008. Post-Earthquake Information Management System (PIMS) Scoping Study. Prepared for the American Lifelines Alliance. September 8. Champaign, IL: University of Illinois.

Porter, K.A., J.A. Beck, and R.V. Shaikhutdinov, 2002. Sensitivity of building loss estimates to major uncertain variables. *Earthquake Spectra* 18: 719–743.

Porter, K.A., S. Hellman, T. McLane, and C. Carlisle, 2010. ROVER: Rapid Observation of Vulnerability and Estimation of Risk. Denver, CO: SPA Risk, LLC. Available at www.sparisk.com/pubs/ATC67-2010-ROVER-flyer.pdf (accessed August 16, 2010).

Pritchard, M.E., 2006. InSAR, a tool for measuring Earth's surface deformation. *Physics Today* (July): 68–69. Available at www.geo.cornell.edu/eas/PeoplePlaces/Faculty/matt/vol59no7p68_69.pdf.

R&C (Rutherford & Chekene Consulting Engineers), 2004. Superior Courts of California, Seismic Assessment Program: Summary Report of Preliminary Findings. Prepared for the California Administrative Office of the Courts, San Francisco, CA. Available at www.courtinfo.ca.gov/reference/documents/seismic0104.pdf (accessed October 6, 2010).

Renschler, C.S., M.W. Doyle, and M. Thoms, 2007. Geomorphology and ecosystems: Challenges and keys for success in bridging disciplines. *Geomorphology* 89: 1–8.

RMS (Risk Management Solutions, Inc.), 2008. 1868 Hayward Earthquake: 140-year retrospective. RMS Special Report. Available at www.rms.com/Publications/1868_Hayward_Earthquake_Retrospective.pdf (accessed August 16, 2010).

Roberts, E.B. and F.P. Ulrich, 1950. Seismological activities of the US coast and geodetic survey in 1948. *Bulletin of the Seismological Society of America* 40: 195–216.

Rogers, E.M., 2003. Diffusion of Innovations (5th Edition). New York: Free Press.

Rose, A., 2002. Model validation in estimating higher-order economic losses from natural hazards. Pp. 105–131 in *Acceptable Risk to Lifeline Systems from Natural Hazard Threats*, C. Taylor and E. Van Marcke (eds.), New York: American Society of Civil Engineers.

Rose, A., 2004. Defining and measuring economic resilience to disasters. *Disaster Prevention and Management* 13: 307–314.

Rose, A., 2005. Analyzing terrorist threats to the economy: A computable general equilibrium approach. Pp. 196–217 in Economic Impacts of Terrorist Attacks, H. Richardson, P. Gordon, J. Moore (eds.), Cheltenham, UK: Edward Elgar.

Rose, A., 2007. Economic resilience to natural and man-made disasters: Multidisciplinary origins and contextual dimensions. *Environmental Hazards* 7: 383–395.

Rose, A., 2009. Economic Resilience to Disasters. Final Report to the Community and Regional Resilience Institute (CARRI). CARRI Research Report 8. Available at www.resilientus.org/library/Research_Report_8_Rose_1258138606.pdf (accessed August 16, 2010).

Rose, A., and S. Liao, 2005. Modeling regional economic resilience to disasters: A computable general equilibrium analysis of a water service disruption. *Journal of Regional Science* 45: 75–112.

Rose, A., and D. Wei, 2007. Indirect loss estimation for water systems. RAMCAP Loss Estimation Software. Washington, DC: ASME Institute.

Rose, A., and T. Szelazek, 2010. An Analysis of the Business Continuity Industry. Center for Risk and Economic Analysis of Terrorism Events (CREATE), University of Southern California, Los Angeles, CA.

Rose, A., G. Oladosu, B. Lee, and G. Beeler-Asay, 2009. The economic impacts of the 2001 terrorist attacks on the World Trade Center: A Computable General Equilibrium Analysis. *Peace Economics, Peace Science, and Public Policy* 15: Article 4.

Rose, A., J. Benavides, S. Chang, P. Szczesniak, and D. Lim, 1997. The regional economic impact of an earthquake: Direct and indirect effects of electricity lifeline disruptions. *Journal of Regional Science* 37: 437–458.

Rose, A., K. Porter, J. Bouabid, C. Huyck, J. Whitehead, D. Shaw, R. Eguchi, T. McLane, L.T. Tobin, P.T. Ganderton, D. Godschalk, A.S. Kiremidjian, K. Tierney, and C.T. West, 2007. Benefit-cost analysis of FEMA hazard mitigation grants. *Natural Hazards Review* 8: 97–111.

Rose, A., S. Liao, and A. Bonneau. In press. Regional economic impacts of a Verdugo earthquake disruption of Los Angeles water supplies: A computable general equilibrium analysis, *Earthquake Spectra*, forthcoming.

Rose, A., D. Wei, and A. Wein. In press. Economic Impacts of the ShakeOut Scenario. *Earthquake Spectra*, forthcoming.

Rowshandel. B., M. Reichle, C. Wills, T. Cao, M. Petersen, D. Branum, and J. Davis., 2003, Estimation of Future Earthquake Losses in California. California Geological Survey.

Rubin, C., 1985. Community Recovery from a Major Natural Disaster. Monograph No. 41: University of Colorado Program on Environment and Behavior, Institute of Behavioral Science.

Schmidtlein, M.C., R.C. Deutsch, W.W. Piegorsch, and S.L. Cutter, 2008. Building indexes of vulnerability: A sensitivity analysis of the social vulnerability index. *Risk Analysis* 28: 1099–1114.

Schwab, J., 1998. Planning for Post-Disaster Recovery and Reconstruction. Planning Advisory Service Report Number 483/484. Chicago, IL: American Planning Association.

Schweitzer, L. 2006. Environmental justice and hazmat transport: A spatial analysis in Southern California. *Transportation Research Part D: Transport and Environment*,11(6): 408–421.

SDR (Subcommittee on Disaster Reduction), 2005. Grand Challenges for Disaster Reduction. National Science and Technology Council (NSTC) Committee on Environment and Natural Resources. Available at www.sdr.gov/GrandChallengesSecondPrinting.pdf (accessed August 13, 2010).

Seligson, H., 2007. HAZUS Modeling for the 1906 San Francisco Earthquake Scenario: Lessons Learned & Suggestions for a New Madrid Earthquake Scenario. Presentation to the New Madrid Earthquake Scenario Workshop, April 20, St. Louis, MO. Available at www.eeri.org/site/images/projects/newmadrid/7-nmes-wshop-sf-scen-seligson.pdf (accessed at August 16, 2010).

Shearer, P., and R. Bürgmann, 2010. Lessons learned from the 2004 Sumatra-Andaman megathrust rupture. *Annual Review of Earth and Planetary Sciences* 38: 103–131.

Shinozuka, M., A. Rose, and R. Eguchi, 1998. Engineering and Socioeconomic Impacts of Earthquakes: An Analysis of Electricity Lifeline Disruptions in the New Madrid Area. MCEER-98-MN02. Buffalo, NY: University of Buffalo.

Smith, V. K., C. Mansfield, and A. Strong, 2008. Can the Economic Value of Security be Measured? Department of Economics, Arizona State University, Phoenix, AZ.

Soong, T. T., S. Y. Chu, and A. M. Reinhorn, 2005. Active, Hybrid, and Semi-active Structural Control: A Design and Implementation Handbook. John Wiley & Sons, Inc.

Spencer, B., and T.T. Soong, 1999. New applications and development of active, semi-active and hybrid control techniques for seismic and non-seismic vibration in the USA. Proceedings of International Post-SMiRT Seminar on Seismic Isolation, Passive Energy Dissipation and Active Vibration of Structures, Cheju, Korea, August 23-25.

SPUR (San Francisco Planning and Urban Research Association), 2009. The Resilient City. Part I: Before the Disaster. San Francisco, CA: SPUR. Available at www.spur.org/publications/library/report/theresilientcity_part1_020109 (accessed August 13, 2010).

Sternberg, E., and K. Tierney, 1998. Planning for Robustness and Resilience: The Northridge and Kobe Earthquakes and Their Implications for Reconstruction and Recovery. Paper presented at US-Japan Workshop on Post-disaster Recovery, Newport Beach, CA, August 24–25.

Swift, J.N., L.L. Turner, J. Benoit, J.C. Stepp, and C.J. Roblee, 2004. Archiving and Web Dissemination of Geotechnical Data: Development of a Pilot Geotechnical Virtual Data Center, Final Report. PEER Lifelines Project 2L02. Berkeley, CA: PEER Lifelines Program, University of California, Berkeley.

Tantala, M., G. Nordenson, G. Deodatis, K. Jacob, B. Swiren, M. Augustyniak, A. Dargush, M. Marrocolo, and D. O'Brien, 2003. Earthquake Risks and Mitigation in the New York, New Jersey, Connecticut Region, NYCEM. Final Summary Report, MCEER -03-SP02. Buffalo, NY: MCEER.

Terra, F.M., I.G. Wong, A. Frankel, D. Bausch, T. Biasco, and J.D. Schelling, 2010. HAZUS analysis of 15 earthquake scenarios in the state of Washington. *Seismological Research Letters* 81: 336.

Thompson, C., 2008. Is the tipping point toast? *Fast Company Magazine*, Issue 122, February. Available at www.fastcompany.com/magazine/122/is-the-tipping-point-toast.html (accessed October 6, 2010).

Tierney, K.J., 1994. Business Vulnerability and Disruption: Data from the Midwest Floods. Paper presented at the 41st North American Meetings of the Regional Sciences Association International, Niagara Falls, Ontario, November 16–20.

Tierney, K.J., 1997. Impacts of recent disasters on businesses: The 1993 Midwest floods and the 1994 Northridge earthquake. Pp. 189–222 in *Economic Consequences of Earthquakes: Preparing for the Unexpected*, B. Jones (ed.), Buffalo, NY: National Center for Earthquake Engineering Research.

Tierney, K.J., 2007. From the margins to the mainstream? Disaster research at the crossroads. *Annual Review of Sociology* 33: 503–525.

Tierney, K.J., M.K. Lindell, and R.W. Perry, 2001. Facing the Unexpected: Disaster Preparedness and Response in the United States. Washington, DC: The Joseph Henry Press.

Tobin, G.A., 1999. Sustainability and community resilience: The holy grails of hazards planning? *Environmental Hazards* 1: 13–25.

UN ISDR (United Nations International Strategy for Disaster Reduction), 2006. Hyogo Framework for Action 2005-2015: Building the Resilience of Nations and Communities to Disasters. Extract from the final report of the World Conference on Disaster Reduction (A/CONF.206/6), March 16, 2005.

URS Corporation, Durham Technologies, Inc., ImageCat, Inc., Pacific Engineering & Analysis, and S&ME, Inc., 2001. Comprehensive Seismic Risk and Vulnerability Study for the State of South Carolina. Final report to the South Carolina Emergency Preparedness Division, Columbia, SC.

U.S.-Canada Power System Outage Task Force, 2004. Final Report on the August 14, 2003 Blackout in the United States and Canada: Causes and Recommendations. Available at reports.energy.gov/ (accessed August 16, 2010).

USGS (United States Geological Survey), 1999. Requirement for an Advanced National Seismic System. Circular 1188. Available at pubs.usgs.gov/circ/c1188/circular.pdf (accessed October 4, 2010).

USGS (United States Geological Survey), 2007. The Plan to Coordinate NEHRP Post-Earthquake Investigations. Prepared in Coordination with the Federal Emergency Management Agency, National Science Foundation, and National Institute of Standards and
Technology. USGS Circular 1242. Available at geopubs.wr.usgs.gov/circular/c1242/
c1242.pdf (accessed August 16, 2010).

Vugrin, E., Warren, D., Ehlen, N., Rose, A., and Barrett, A, 2009. Chemical Supply Chain
and Resilience Project: A Resilience Definition for Use in Economic and Critical Infrastructure Resilience Analysis. Prepared for the U.S. Department of Homeland Security
Science and Technology Directorate, Sandia National Laboratories, Albuquerque, NM
and National Center for Risk and Economic Analysis of Terrorism Events (CREATE),
University of Southern California, Los Angeles, CA, August 24.

Wald, D. J., B. Worden, V. Quitoriano, and J. Goltz, 2001. Practical Applications for Earthquake Scenarios Using ShakeMap. American Geophysical Union, Fall Meeting 2001,
abstract #S32D-01.

Weaver, C., B.L. Sherrod, R.A. Haugerud, K.L. Meagher, A.D. Franke, S.P. Palmer, and R.J.
Blakely, 2005. The scenario earthquake and ground motions. Chapter 1 in M. Stewart,
(ed.), Scenario for a Magnitude 6.7 Earthquake on the Seattle Fault. Oakland, CA, EERI,
and Camp Murray, WA, Washington Military Department, Emergency Management
Division.

Webb, G.R., K.J. Tierney, and J.M. Dahlhamer, 2000. Business and disasters: Empirical
patterns and unanswered questions. *Natural Hazards Review* 1: 83–90.

Whitehead, J., S. Pattanayek, B. Houtven, and B. Gelso, 2008. Combining revealed and stated
preference data to estimate the nonmarket value of ecological services: An assessment
of the state of the science. *Journal of Economic Surveys* 22: 874–908.

Whitehead, J., and A. Rose, 2009. Estimating environmental benefits of natural hazard mitigation: Results from a benefit-cost analysis of FEMA mitigation grants. *Mitigation and
Adaptation Strategies for Global Change* 14: 655–676.

Whittaker, A.S., and R.C. Krumme, 1993. Structural Control Using Shapememory Alloys.
E*Sorb Systems Report No. 9301. Berkeley, CA.

Williams, M.L., K.M. Fischer, J.T. Freymueller, B. Tikoff, A.M. Trehu, et al., 2010. Unlocking
the Secrets of the North American Continent: An EarthScope Science Plan for 2010-
2020, February, 2010. 78 pp. Available at www.earthscope.org/ESSP (accessed October
4, 2010).

WGCEP (Working Group on California Earthquake Probabilities), 2008. The Uniform
California Earthquake Rupture Forecast, Version 2 (UCERF 2), U.S. Geological Survey
Open File Report 2007-1437. Reston, VA.

Zechar, J.D., D. Schorlemmer, M. Liukis, J. Yu, F. Euchner, P.J. Maechling, and T. H. Jordan,
2009. The Collaboratory for the Study of Earthquake Predictability perspectives on
computational earth science, *Concurrency & Computation* 22: 1836–1847.

Zhang, Y., and W.G. Peacock, 2010. Planning for Housing Recovery? Lessons Learned From
Hurricane Andrew. *Journal of the American Planning Association* 76: 1, 5–24.

Zielke, O., J. R. Arrowsmith, L. G. Ludwig, and S. O. Akciz, 2010. Slip in the 1857 and
Earlier Large Earthquakes Along the Carrizo Plain, San Andreas Fault. DOI:10.1126/
science.1182781. *Science* 327: 1119–1122.

Zoback, M.L., and P. Grossi, 2010. The next Hayward earthquake—who will pay? Pp. XXX–XXX
in K. Knudsen, J. Baldwin, T. Brocher, R. Burgmann, M. Craig, D. Cushing, P. Hellweg,
M. Wiegers, and I. Wog, (eds.), *Proceedings of the Third Conference on Earthquake Hazards in
the eastern San Francisco Bay Area*. California Geological Survey, Special Publication xxx.

Appendixes

Appendix A

Summary of 2008 NEHRP Strategic Plan

The Strategic Plan for the National Earthquake Hazards Reduction Program (NEHRP) for Fiscal Years 2009-2013 was submitted to Congress by the Interagency Coordinating Committee (ICC) of NEHRP, as required by the Earthquake Hazards Reduction Act of 1977. The plan outlines a cooperative program of earthquake monitoring, research, implementation, education, and outreach activities to be performed by the NEHRP agencies—the Federal Emergency Management Agency, the National Institute of Standards and Technology (NEHRP lead agency), the National Science Foundation, and the U.S. Geological Survey.

The plan is founded on the premise that the continued success of NEHRP will emphasize the linked roles of the NEHRP agencies and their partners, based on a common vision and shared mission. The vision is "a nation that is earthquake-resilient in public safety, economic strength, and national security;" the mission is:

> To develop, disseminate, and promote knowledge, tools, and practices for earthquake risk reduction—through coordinated, multidisciplinary, interagency partnerships among the NEHRP agencies and their stakeholders—that improve the Nation's earthquake resilience in public safety, economic strength, and national security.

Accomplishing the NEHRP mission will require developing and applying knowledge based on research in the geological, engineering, and social sciences; educating leaders and the public; and assisting state, local, and private-sector leaders to develop standards, policies, and practices. The NEHRP agencies have established three overarching, long-term Strategic Goals, with 14 associated objectives, to support this mission:

Goal A: Improve understanding of earthquake processes and impacts.

- Objective 1: Advance understanding of earthquake phenomena and generation processes.
- Objective 2: Advance understanding of earthquake effects on the built environment.
- Objective 3: Advance understanding of the social, behavioral, and economic factors linked to implementing risk reduction and mitigation strategies in the public and private sectors.
- Objective 4: Improve post-earthquake information acquisition and management.

Goal B: Develop cost-effective measures to reduce earthquake impacts on individuals, the built environment, and society-at-large.

- Objective 5: Assess earthquake hazards for research and practical application.
- Objective 6: Develop advanced loss estimation and risk assessment tools.
- Objective 7: Develop tools that improve the seismic performance of buildings and other structures.
- Objective 8: Develop tools that improve the seismic performance of critical infrastructure.

Goal C: Improve the earthquake resilience of communities nationwide.

- Objective 9: Improve the accuracy, timeliness, and content of earthquake information products.
- Objective 10: Develop comprehensive earthquake risk scenarios and risk assessments.
- Objective 11: Support development of seismic standards and building codes and advocate their adoption and enforcement.
- Objective 12: Promote the implementation of earthquake-resilient measures in professional practice and in private and public policies.
- Objective 13: Increase public awareness of earthquake hazards and risks.
- Objective 14: Develop the nation's human resource base in earthquake safety fields.

The plan also describes nine cross-cutting Strategic Priorities that directly support the goals and augment other on-going agency activities needed to satisfy them. These priorities are:

- Fully implement the Advanced National Seismic System.
- Improve techniques for evaluating and rehabilitating existing buildings.
- Further develop Performance-Based Seismic Design.
- Increase consideration of socioeconomic issues related to hazard mitigation implementation.
- Develop a national post-earthquake information management system.
- Develop advanced earthquake risk mitigation technologies and practices.
- Develop guidelines for earthquake-resilient lifeline components and systems.
- Develop and conduct earthquake scenarios for effective earthquake risk reduction and response and recovery planning.
- Facilitate improved earthquake mitigation at state and local levels.

The goals, objectives, and Strategic Priorities are consistent with, and expand upon, the "Grand Challenges for Disaster Reduction: Priority Interagency Earthquake Implementation Actions" identified by the Subcommittee on Disaster Reduction of the President's National Science and Technology Council.

The plan provides a straightforward and executable strategy for NEHRP. Successful strategic planning and program accomplishment must be consistent with existing policies, based on realistic assumptions, and responsive to changing conditions. The pace of program accomplishment will depend on the resources that are available to NEHRP agencies during the 2009-2013 plan period, and the plan is intended to guide relevant funding decisions by NEHRP agencies. Following the adoption of the plan, the NEHRP agencies propose to jointly develop a Management Plan to detail Strategic Plan implementation activities that are consistent with agency appropriations and funding priorities.

The costs of earthquake loss reduction and post-earthquake recovery are shared by the public and private sectors. The role of NEHRP is to provide the public and private sectors with the scientific and engineering information, knowledge, and technologies needed to prepare for earthquakes and thus reduce the costs of losses and recovery. NEHRP will continue to develop partnerships with its stakeholder community of earthquake professionals working in academia and in business, government, technical, professional, and codes-and-standards organizations that are involved with the earthquake risk reduction process, in fulfillment of its role.

The NEHRP agencies propose to keep abreast of advances in science and technology, and adjust short- and long-term developmental efforts accordingly. Although NEHRP will remain focused on the elements of the Strategic Plan, the agencies will adapt to contingencies and opportunities that may arise. If a major earthquake occurs in the United States during the plan period, NEHRP will initiate efforts to study the effects and impacts of that event, including successes, failures, and unforeseen problems that arose in mitigation, response, and recovery practices and policies, and adjust the plan as needed.

Appendix B

Summary of 2003 EERI Report

In 2003 the Earthquake Engineering Research Institute (EERI) released the report, *Securing Society against Catastrophic Earthquake Losses: A Research and Outreach Plan in Earthquake Engineering*. This report was prepared by a panel of earth scientists, earthquake engineers, and social scientists involved in earthquake-related research, with input from professional communities throughout the United States, and its goal was to provide a vision for the future of earthquake engineering research and outreach focused on securing the nation against the catastrophic effects of earthquakes.

The plan comprised the following research and outreach programs:

- *Understanding Seismic Hazards*: developing new models of earthquakes and seismic hazards based on fundamental physics.
- *Assessing Earthquake Impacts*: evaluating the impact of disasters on the built environment by simulating performance of structures and entire urban systems.
- *Reducing Earthquake Impacts*: developing new materials, structural and nonstructural systems, lifeline systems, foundation systems, tsunami protection, fire protection systems, and land-use measures.
- *Enhancing Community Resilience*: exploring new ways to reduce risk and improve the decision-making capability of stakeholders.
- *Expanding Education and Public Outreach*: improving the education of engineers and scientists from elementary school to advanced graduate education, and providing opportunities for the public to learn about earthquake risk reduction.

The research tasks for each program were intended to develop the science, engineering, and societal approaches necessary for making better risk management choices to prevent catastrophic losses. The outreach tasks, on the other hand, were intended to facilitate the transfer of research findings into practice. The report proposes that achieving the goal of catastrophic loss prevention requires not only technological breakthroughs but also the translation of research results into professional practice and decision-making. For example, the report identified one central focus of earthquake engineering research as the need to merge current and future information technology advances into the practice of earthquake engineering, with the objective of reducing the uncertainty associated with hazard, performance, damage, and loss prediction of the built environment. However, while loss-reduction strategies that address specific structures and systems are important, the plan also stressed the need to protect the social fabric of communities against earthquake losses, requiring more comprehensive and holistic approaches.

The cost of the plan was estimated at $358 million per year for the first 5 years of a 20-year program of funding for activities within the NEHRP agencies. The total estimate for the 20-year plan, including capital investments, was $6.54 billion, with the expectation that funds would ramp up at a 15 percent annual rate over the first 5-year period of the Plan. Details of the budget over the 20-year period are presented in Tables B.1 and B.2.

The report indicated that accomplishing the plan would require a high level of coordination among the NEHRP agencies, as well as with other federal agencies and state and local government organizations, the earthquake engineering research community, organizations responsible for promulgation of building codes, engineering professionals, and government officials. Importantly, the benefits would not be limited to preventing catastrophic losses from earthquakes. Plan outcomes would also provide substantial benefits for homeland security and other initiatives to increase community resilience to extreme events. Through advances in the design of buildings and facilities, planning measures for addressing population growth and land use, and technologies that address emergency management and recovery, the initiatives presented in the report would complement and enhance programs to reduce the threat of terrorist attack and harmful effects of other extreme events such as blast, wind, flood, and fire.

TABLE B.1 Estimated Cost of Plan (millions$), Including Research and Outreach Programs and Related Activities

Activity	Average Annual Cost (M$)				Total 20-year Cost (M$)
	FY04-08	FY09-13	FY14-18	FY19-23	
Hazard Knowledge	86	86	70	55	1,485
Impact Assessment	64	67	36	21	940
Impact Reduction	82	92	60	41	1,375
Enhancing Community Resilience	22	33	44	44	715
Education and Public Outreach	20	20	20	20	400
Capital Investments	55	77	80	70	1,410
Information Technology	28	5	5	5	215
Management Plan Development	1	0	0	0	5
PLAN TOTAL	358	380	315	256	6,545

NOTE: Capital Investments include ANSS, NEES, and field instrumentation.

TABLE B.2 Distribution of Costs (millions$) Among Research, Education, and Outreach Programs, Capital Investment, Information Technology, and Program Management

Program Description		Average Annual Cost (M$)				Total 20-year Cost (M$)
		FY04-08	FY09-13	FY04-08	FY09-13	
Hazard Knowledge	Research	36	36	30	25	635
	Outreach	50	50	40	30	850
Impact Assessment	Research	61	61	30	15	835
	Outreach	3	6	6	6	105
Impact Reduction	Research	64	65	38	24	955
	Outreach	18	27	22	17	420
Community Resilience	Research	10	15	20	20	325
	Outreach	12	18	24	24	390
Education/Public Outreach		20	20	20	20	400
Capital Investments		55	77	80	70	1,410
Information Technology		28	5	5	5	215
Management Plan Development		1	0	0	0	5
PLAN TOTAL		358	380	315	256	6,545

Appendix C

Committee and Staff Biographies

Robert M. Hamilton (*Chair*) is a seismologist with a primary interest in natural disaster loss reduction. He retired as Deputy Executive Director of NRC's Division on Earth and Life Studies in 2004. He had previously served as Executive Director of NRC's Commission on Geosciences, Environment, and Resources, following 30 years as a geophysicist with the U.S. Geological Survey. He chaired the Committee on Disaster Reduction for the International Council for Science (ICSU), and chaired the Scientific and Technical Committee of the International Decade for Natural Disaster Reduction (IDNDR), a United Nations program for the 1990s. He also served for 2 years with the IDNDR Secretariat in Geneva, including a year as Director. He has been a member of the Inter-agency Task Force for the International Strategy for Disaster Reduction, a follow-on United Nations program to the IDNDR. He also chaired the Subcommittee on Disaster Reduction of the National Science and Technology Council. Dr. Hamilton served as President of the Seismological Society of America, and President and Secretary of the Seismology Section of the American Geophysical Union. He is a Fellow of the Geological Society of America and the American Association for the Advancement of Science. Dr. Hamilton has a geophysical engineering degree from Colorado School of Mines, and M.A. and Ph.D. degrees in geophysics from the University of California, Berkeley.

Richard A. Andrews has more than 30 years' experience in emergency management, counter-terrorism policy, and seismic safety. He is a member of the Homeland Security Advisory Council, which provides policy guidance to the Department of Homeland Security (DHS) and the Federal Emergency Management Agency (FEMA) National Advisory Council. He chairs the Council's Senior Advisory Committee on Emergency Services, Law Enforcement, Public Health and Hospitals. He served as Director of the California Office of Homeland Security and Homeland Security Advisor to Governor Arnold Schwarzenegger from 2004-2005. From 1991 to 1998, Dr. Andrews was Director of the Governor's Office of Emergency Services for California, where he managed the emergency response and recovery efforts for 19 presidential and 24 gubernatorial disasters. He is a member of the World Bank's Disaster Management Operations Group and has worked on emergency management projects in Turkey, Algeria, Romania, and India. Dr. Andrews is a past President of the National Emergency Management Association (NEMA) and former Executive Director of the California Seismic Safety Commission. He is the former Chair of NEMA's Private Sector Committee as well as a public-private task force formed to explore ways in which the Emergency Management Assistance Compact—a congressionally ratified organization that provides form and structure to interstate mutual aid—might be employed to more effectively use private-sector resources during major emergencies. Dr. Andrews received an A.B. from DePauw University, and an M.A. and Ph.D. from Northwestern University.

Robert A. Bauer is an engineering geologist and head of the Engineering and Coastal Geology Section of the Illinois State Geological Survey (ISGS). He has worked with Illinois state emergency managers on exercises and workshops since 1990 and is the ISGS/Institute of Natural Resource Sustainability (INRS) representative to the state's emergency operations center. He has participated on the earthquake scenario committees and hazard map production for the Illinois statewide earthquake assessment. He is the ISGS/INRS' representative and State Geologist Technical Director, program coordinator, and past-Chair of the Association of the Central U.S. Earthquake Consortium State Geologists. He serves on the Illinois Seismic Safety Task Force, Earthquake Engineering Research Institute (EERI) New Madrid Scenario Executive Committee, and provided important input to the FEMA New Madrid Catastrophic Planning Scenario subcommittee. He has authored more than 90 publications, and is a member of the Geo-Institute of ASCE, Association of Engineering Geologists, Society of Mining Engineers of the American Institute of Mining, Metallurgical, and Petroleum Engineers, International Association of Engineering Geologists, and EERI. Mr. Bauer received a B.S. in geological science from the University of Illinois at Chicago, and an M.S. in engineering geology from the University of Illinois at Urbana-Champaign.

Jane A. Bullock is a principal at Bullock and Haddow, LLC, a disaster mitigation consulting firm, and also is an adjunct professor at the Institute for Crisis, Disaster, and Risk Management at George Washington University. Ms. Bullock has more than 25 years of private- and public-sector experience culminating in responsibility, as chief of staff, for the daily management and operations of FEMA, with its responsibility for disaster mitigation, response, and recovery. In the course of her career, she directed the restructuring and streamlining of the agency, set policy and programmatic direction for the nation's emergency management systems, served as the agency's spokesperson, and worked with Congress and the nation's governors to enhance disaster management throughout the United States. She was chief architect of Project Impact: Building Disaster Resistant Communities, a nationwide, grassroots effort by communities and businesses to implement mitigation and risk reduction programs. In 2000, she received the Presidential Rank Award, the highest award presented by the President to a career civil servant. Since leaving FEMA, Ms. Bullock has worked with a variety of organizations to design and implement disaster management and homeland security programs. In the post-Katrina environment, she has worked with Save the Children to design and implement their domestic disaster response and recovery program. She testified before both House and Senate committees about the future of emergency management after Hurricane Katrina. Internationally, she has worked with countries in Central and South America, Eastern Europe, and New Zealand on implementing disaster management and mitigation programs. She is the coauthor of textbooks on emergency management, homeland security, climate change and mitigation, and a *Living with the Shore* book series dealing with the design and construction of communities in hazardous areas.

Stephanie E. Chang is a professor at the University of British Columbia (UBC), where she has joint faculty appointments with the School of Community and Regional Planning and the Institute for Resources, Environment, and Sustainability. She holds a Canada Research Chair position (tier 2) in Disaster Management and Urban Sustainability. Much of Dr. Chang's work aims to bridge the gap between engineering, natural sciences, and social sciences in addressing the complex issues of natural disasters. Some of her research has focused on developing integrated regional models for estimating losses from future earthquakes. She has also developed methods for assessing disaster mitigation strategies and researched how disasters impact regional economies. Her current research addresses community disaster resilience and sustainability, mitigation of infrastructure system risks (especially electric power, water, and transportation), and urban disaster recovery. She is particularly interested in applications to cities of the Pacific Rim. Prior to joining UBC, she was a research assistant professor in the Department of Geography at the University of Washington. She has also worked as a researcher and consultant with EQE International (subsequently ABS Consulting) in Los Angeles and Seattle. Dr. Chang was awarded the 2001 Shah Family Innovation Prize by EERI and served on the editorial board of *Earthquake Spectra*. She recently served on the National Research Council's Committee on Disaster Research in the Social Sciences. Dr. Chang received a B.S.E. in civil engineering and operations research from Princeton University, and an M.S. and Ph.D. in regional science from Cornell University.

William T. Holmes is a vice president and structural engineer at Rutherford and Chekene, Consulting Engineers, a multi-disciplinary engineering firm. Mr. Holmes has 40 years of practical experience in all aspects of designing structures, particularly design for protection from earthquake effects. In addition to traditional structural engineering design of buildings, Mr. Holmes' broad interests and experience include post-earthquake reconnaissance and analysis, post-earthquake response of hospitals, seismic protection of nonstructural systems, fragility and retrofit standards for unreinforced masonry buildings, regional loss estimation, development of seismic standards for both new and existing buildings, research and development of seismic technology, seismic isolation, public policy, and performance-based seismic engineering. Mr. Holmes has traveled to Armenia, Azerbaijan, Canada, China, Ecuador, Greece, India, Italy, Japan, Mexico, New Zealand, Pakistan, Thailand, and Turkey to address conferences and workshops or to consult with local officials on seismic design. As a result of his long and varied career, he has been awarded the Alfred E. Alquist Medal for Achievement in Earthquake Safety (Public Service) by the California Earthquake Safety Foundation, the H.J. Brunnier Award for lifetime achievement in structural engineering by the Structural Engineers Association of Northern California (SEAONC), the Exceptional Service Award by the Building Seismic Safety Council, and Honorary Membership in the Structural Engineers Association of California and the Earthquake Engineering Research Institute (EERI). He sits on the Board of Directors of Consortium of Universities for Research in Earthquake Engineering (CUREE) and has served as President of SEAONC and the Applied Technology Council. Mr. Holmes received a B.S. in civil engineering and an M.S. in structural engineering from Stanford University.

Laurie A. Johnson is Principal of Laurie Johnson Consulting and Research. She has more than 20 years of professional experience in urban planning, risk management, and disaster recovery research and consulting. She has written extensively about the economics of catastrophes, land use and risk, and urban disaster recovery and reconstruction, and researched most of the large-scale urban disasters of the past 20 years, including the 2008 Sichuan China earthquake, Hurricane Katrina, the 2001 World Trade Center collapse, and the 1994 Northridge, CA, and 1995 Kobe, Japan, earthquakes. In March 2006, she founded her consultancy, working to apply the principles and technologies of risk management to solve complex urban problems. Her clients include the California Governor's Office of Emergency Services, Fritz Institute, Greater New Orleans Community Support Foundation, and the U.S. Geological Survey. In 2006 and 2007, she was a lead author and disaster recovery expert on the development of a unified recovery and rebuilding plan for the City of New Orleans following the devastation of Hurricane Katrina. She is also an International Research Collaborator at the Research Center for Disaster Reduction Systems at the Disaster Prevention Research Institute. She is on the Board of Directors of the Public Entity Risk Institute, and a member of the Earthquake Engineering Research Institute, American Institute of Certified Planners, and the American Planning Association. She holds a master of urban planning and B.S. degrees, both from Texas A&M University, and a doctorate of informatics from Kyoto University, Japan.

Thomas H. Jordan (NAS) is director of the Southern California Earthquake Center (SCEC) and W. M. Keck Professor of Earth Sciences at the University of Southern California (USC). He oversees all aspects of SCEC's program, which currently involves more than 600 scientists at more than 60 universities and research institutions. SCEC develops comprehensive understanding of earthquakes and communicates knowledge for reducing earthquake risk. Dr. Jordan is a member of the California Earthquake Prediction Evaluation Council and the NAS Council and the NRC Governing Board. His research addresses earthquake processes, seismology of the earth, and geodetic observations of plate motions and interplate deformation. His other areas of interest include continental formation and tectonic evolution, mantle dynamics, and statistical descriptions of seafloor morphology. Dr. Jordan is the author or co-author of approximately 180 scientific publications, including the NRC decadal report, *Living on an Active Earth: Perspectives on Earthquake Science*, and two popular textbooks. He taught at Princeton University and the Scripps Institution of Oceanography before joining the Massachusetts Institute of Technology (MIT) as the Robert R. Shrock Professor in 1984. He served as the head of MIT's Department of Earth, Atmospheric and Planetary Sciences for the decade 1988-1998. In 2000, he moved from MIT to USC. He has been awarded the Macelwane and Lehmann Medals of AGU and the Woollard Award of GSA. He has been elected to NAS, the American Academy of Arts and Sciences, and the

American Philosophical Society. Dr. Jordan received his B.A., M.S., and Ph.D. from the California Institute of Technology.

Gary A. Kreps is professor emeritus and former Vice Provost at the College of William and Mary. He began his career as a faculty member and administrator at William and Mary and continued there until retiring in July 2005. Dr. Kreps has long-standing research interests in organizational and role theories as both relate to structural analyses of community, regional, and societal responses to natural, technological, and willful hazards and disasters. He has served as a staff member, consultant, or member on five National Research Council committees: the Committee on the Socioeconomic Effects of Earthquake Prediction, the Committee on U.S. Emergency Preparedness, the Committee on International Disaster Assistance, the Committee on Mass Media Reporting of Disasters, and the Committee on Disaster Research in the Social Sciences. Over the course of the past 2 decades, Dr. Kreps and his collaborators have developed taxonomies and theories of organizing and role enactment during the emergency periods of disasters. Major findings from his research program have been reported in two books and articles in *Sociological Theory, Annual Review of Sociology, American Sociological Review, American Journal of Sociology, Journal of Applied Behavioral Science, International Journal of Mass Emergencies and Disasters*, and many other basic and applied publications. Dr. Kreps' 2001 entry in the *International Encyclopedia of the Social and Behavioral Sciences* ("Disaster, Sociology of") emphasizes the need to reconcile functionalist and constructivist conceptions of disasters as acute systemic events. Most recently, he received the 2008 E.L. Quarantelli Award for career contributions to social science theory and research on hazards and disasters. Dr. Kreps received his bachelor's degree in sociology at the University of Akron and his master's and doctorate degrees from The Ohio State University.

Stuart Nishenko is the Senior Seismologist in the Geosciences Department of the Pacific Gas and Electric Company in San Francisco, CA. His focus is on earthquake hazard assessment and risk management, and he has authored or co-authored more than 100 publications including the 2001 FEMA 366 *HAZUS99 Estimated Annualized Earthquake Losses for the United States* study, the 1988 and 1990 *Working Group on California Earthquake Probability* reports, and 2006 NRC study on the *Economic Benefits of Improved Seismic Monitoring*. He serves as a member the USGS Scientific Earthquake Studies Advisory Committee, and as chairman of the California Integrated Seismic Network Advisory Committee and the Government Relations Committee of the Seismological Society of America. He received his Ph.D. in geophysics from Columbia University, Lamont-Doherty Earth Observatory in 1983 and was a NRC Postdoctoral Research Associate.

Dr. Nishenko is the liaison to this committee from the Committee on Seismology and Geodynamics.

Adam Z. Rose is a research professor at the University of Southern California School of Policy, Planning, and Development. He is also Coordinator for Economics at USC's DHS Center for Risk and Economic Analysis of Terrorism Events. Much of Dr. Rose's research is on the economics of natural and man-made hazards. He recently served on an NRC panel on the economic benefits of seismic monitoring, as a lead researcher for a report to Congress on the net benefits of FEMA hazard mitigation grants, as lead economist on the Southern California ShakeOut Project, as co-principal investigator (PI) on a study to develop a hazards decision-support system for the Los Angeles Department of Water and Power, and as coordinator for the DHS Integrated Network of Centers of a set of studies on economic and community resilience. He is currently a co-PI on an NSF grant to estimate the economic impacts of risk amplification following terrorist attacks. A major focus of his research has been on resilience to natural disasters and terrorism at the levels of the individual business, market, and regional economy. Dr. Rose's other research areas are the economics of energy and climate change policy. He has served on the editorial boards of the *Journal of Regional Science*, *Resource and Energy Economics*, *Energy Policy*, and *Resource Policy*. He has served as the American Economic Association Representative to the American Association for the Advancement of Science (AAAS), and on the Board of Directors of the American Association of Geographers Energy and Environment Specialty Group. He is the recipient of a Woodrow Wilson Fellowship, East-West Center Fellowship, American Planning Association's Outstanding Program Planning Honor Award, EERI Special Service Recognition Award, and Applied Technology Council Outstanding Achievement Award. Dr. Rose received a B.A. in economics from the University of Utah, and an M.A. and Ph.D. in economics from Cornell University.

L. Thomas Tobin is a consultant with Tobin & Associates. He has worked on natural hazards, risk management, and public policy issues for 40 years. Mr. Tobin served 10 years as Executive Director of the California Seismic Safety Commission. He has lobbied for legislation, having testified to Congressional committees on six occasions and state legislative committees on more than 100 occasions. He served on the NEHRP advisory committee from 1991 to 1993 and the California State Historical Building Safety Board from 1991 to 1995. He served as a Director and Vice President of EERI, was EERI's Distinguished Lecturer in 1996, and was presented the San Jose State University College of Engineering's Award of Distinction in 1996. He was the 2004 recipient of the Alfred E.

Alquist Medal for Achievement in Earthquake Safety. He was the found-ing Secretary-Treasurer of EERI's northern California Chapter from 2001 through 2003, and is the current President. As a consultant, Mr. Tobin helped FEMA create both Project Impact and the Disaster Resistant University initiatives. He currently is involved in projects advocating earthquake resilience and mitigation through land-use regulation and planning and by integrating seismic safety principles with his clients' ongoing activities. He is senior advisor at GeoHazards International, bringing resources and technical knowledge to developing countries to reduce earthquake risk, and vice chair of the Multihazard Mitigation Council. He is a registered professional engineer. Mr. Tobin received a B.S. in civil engineering from the University of California at Berkeley, and an M.S. in geotechnical engineering from San José State University.

Andrew S. Whittaker is a professor and department chair in the Depart-ment of Civil, Structural and Environmental Engineering at the University at Buffalo, State University of New York, and a licensed structural engineer in the state of California. He practiced as a structural engineer in Australia and Asia in the late 1970s and early 1980s and in the United States in the late 1980s. He served as the associate director of the Earthquake Engineering Research Center and Pacific Earthquake Engineering Research Center in the 1990s and joined the University at Buffalo in 2000. He joined the Board of Directors of CUREE in 2001, served as Vice President in 2003-2004, and has been President since 2005. Dr. Whittaker's research and professional inter-ests include earthquake and blast engineering, performance-based design, seismic protective systems, ultra-high-rise buildings, offshore platforms, and power-related infrastructure. He is the author of more than 200 publica-tions, including a reference text, book chapters, journal papers, conference papers, and technical reports. Dr. Whittaker led NSF-funded earthquake reconnaissance teams to Kobe, Japan, in 1995, and Izmit, Turkey, in 1999, and was a member of the three-person, NSF-funded structural engineer-ing reconnaissance team at the site of the former World Trade Center in September 2001. He currently serves on technical committees for American Concrete Institute, American Society of Civil Engineers (ASCE), American Institute of Steel Construction, Building Seismic Safety Council, FEMA, EERI, and USGS. Dr. Whittaker provides consulting and peer-review ser-vices to private companies, local, state, and federal government agencies in the United States, Asia, Australia, Europe, Far East, Middle East, South America, and the United Kingdom. A focus of his professional work is the application of new technologies and performance-based design to ultra-tall buildings, bridges, and conventional and nuclear-related infrastructure. He is the leader for the Structural Performance Products team that is develop-ing the second generation of tools for performance-based earthquake engineering as part of the DHS/FEMA-funded ATC-58 (Applied Technology Council 58) project. Dr. Whittaker received a B.E. in civil engineering from the University of Melbourne, Australia, and a M.S. in civil engineering and Ph.D. in structural engineering from the University of California at Berkeley.

NATIONAL RESEARCH COUNCIL STAFF

David A. Feary is a Senior Program Officer with the NRC's Board on Earth Sciences and Resources and Staff Director of BESR's Committee on Seismology and Geodynamics. He is also a research professor in the School of Earth and Space Exploration and the School of Sustainability at Arizona State University. Prior to joining the NRC, he spent 15 years as a research scientist with the marine program at Geoscience Australia. During this time, he participated in numerous national and international research cruises to better understand the role of climate as a primary control on carbonate reef formation and to improve understanding of cool-water carbonate depositional processes and controls. He is a member of the Science Planning Committee of the Integrated Ocean Drilling Program. Dr. Feary received B.Sc. and M.Sc. (Hons) degrees from the University of Auckland and his Ph.D. from the Australian National University.

Appendix D

Community Workshop Attendees and Presentations to Committee

WORKSHOP ATTENDEES

Walter Arabasz
University of Utah

Ralph Archuleta
University of California, Santa Barbara

Mark Benthien
University of Southern California

Jonathan Bray
University of California, Berkeley

Arrietta Chakos
Harvard Kennedy School, Taubman Center for State and Local
    Government

Mary Comerio
University of California, Berkeley

Reginald DesRoches
Department of Civil Engineering, Georgia Tech

Andrea Donnellan
National Aeronautics and Space Administration

Leonardo Duenas-Osorio
Rice University

Paul Earle
U.S. Geological Survey

Richard Eisner
Fritz Institute

Ronald Eguchi
Imagecat, Inc.

Art Frankel
U.S. Geological Survey

James Goltz
California Governor's Office of Emergency Services

Ronald Hamburger
Simpson Gumpertz & Heger

Jim Harris
J. R. Harris & Company

Jack Hayes
National Earthquake Hazards Reduction Program, National Institute of
    Standards and Technology

Jon Heintz
Applied Technology Council

Eric Holdeman
Eric Holdeman & Associates

Doug Honegger
D.G. Honegger Consulting

Richard Howe
R.W. Howe & Associates, PLC

Theresa Jefferson
Center for Technology, Security, and Policy, Virginia Polytechnic
    Institute and State University

Lucy Jones
U.S. Geological Survey

Michael Lindell
Texas A&M University

Nicolas Luco
U.S. Geological Survey

Steven Mahin
Pacific Earthquake Engineering Research Center, University of
California, Berkeley

Peter May
Center for American Politics and Public Policy, Political Science
Department, University of Washington

Dick McCarthy
California Seismic Safety Commission

David Mendonça
New Jersey's Science & Technology University

Dennis Mileti
Natural Hazards Center

Robert Olson
Robert Olson Associates, Inc.

Chris Poland
Degenkolb Engineers

Woody Savage
U.S. Geological Survey

Hope Seligson
MMI Engineering

Kimberley Shoaf
University of California, Los Angeles

Paul Somerville
URS Corporation

Kathleen Tierney
Natural Hazards Center, University of Colorado

Susan Tubbesing
Earthquake Engineering Research Institute

John Vidale
Pacific Northwest Seismic Network

Yumei Wang
Oregon Dept of Geology and Mineral Industries

Gary Webb
Oklahoma State University

Sharon Wood
Department of Civil, Architectural and Environmental Engineering,
    University of Texas at Austin

Brent Woodworth
Los Angeles Emergency Preparedness Foundation

Mary Lou Zoback
Risk Management Solutions, Inc.

WORKSHOP BREAKOUT QUESTIONS

Cross-disciplinary Breakout—
Elements of an Earthquake-Resilient Nation

1: What does an earthquake-resilient nation look like?
2: How should we measure national resilience? How would we know if we are becoming more earthquake resilient?

Disciplinary Breakout—
Fundamental Science Requirements and Enabling Technologies

3: What are the fundamental science requirements in your disciplinary area for an earthquake-resilient nation?
4: What are the enabling technologies in your disciplinary area needed for an earthquake-resilient nation?

Disciplinary Breakout—Implementation and Policy Aspects

5: What are the implementation challenges and opportunities needed for an earthquake-resilient nation?
6: What are the behavioral changes needed for an earthquake-resilient nation?

Cross-disciplinary Breakout—Fundamental Science Requirements

7: Are there any missing activities contained in Question 3 responses?
8: What activities are critical for advancing resilience?
9: What is the sequence of activities?

Cross-disciplinary Breakout—Enabling Technologies

10: Are there any missing activities contained in Question 4 responses?
11: What activities are critical for advancing resilience?
12: What is the sequence of activities?

Cross-disciplinary Breakout—Implementation and Policy Aspects

13: Are there any missing activities contained in Questions 5 and 6 responses?
14: What activities are critical for advancing resilience?
15: What is the sequence of activities?

PRESENTATIONS TO THE COMMITTEE

Meeting 1:

Context of the Project   *Shyam Sunder, NIST*

Sponsor Hopes and Expectations   *Jack Hayes, NEHRP/NIST*

NEHRP Strategic Plan   *Jack Hayes, NEHRP/NIST*

FEMA Update—Earthquake Resilience Activities   *Ed Laatsch, FEMA*
                                                                 Mike Mahoney, FEMA

Role of the U.S. Geological Survey in the
National Earthquake Hazards Reduction Program
                                                              David Applegate, USGS

Role of the National Science Foundation in the
National Earthquake Hazards Reduction Program
                                                            Joy Pauschke, NSF-ENG
                                                            Eva Zanzerkia, NSF-GEO
                                                        Richard Fragaszy, NSF-ENG
                                                          Dennis Wenger, NSF-ENG

Meeting 2:

Cost Estimates in the EERI "Securing Society" Report
                                                   Paul Somerville, URS Corporation

Meeting 3:

Closed session only.

Meeting 4:

Closing comments from the NSF   *Joy Pauschke, NSF-ENG*

Closing comments from NIST/NEHRP   *Jack Hayes, NEHRP/NIST*

Closing comments from the U.S. Geological Survey   *David Applegate*

Closing comments from the Federal Emergency Management Agency
                                                                 Mike Mahoney, FEMA

Status of NEHRP reauthorization process   *Jack Hayes, NEHRP/NIST*

An Individual Perspective based on extensive involvement with NEHRP
John Filson, USGS Emeritus

Appendix E

Additional Cost Information

TABLE E.1 Detailed Task Breakdown (in $000) and Scheduling for Task 4: National Seismic Hazard Model

Task Component	Year	Task Breakdown ($)	Years 1-5 (Annualized) ($)	Years 6-10 (Annualized) ($)	Years 11-20 (Annualized) ($)
Provide geologic information to map faults and provide data for predictive relationships	1-20	560,000	28,000	28,000	28,000
USGS design applets and website updates	1-20	10,000	500	500	500
Update predictive relationships for ground shaking, including earthquake-physics simulations	1-20	153,000	10,200	6,400	7,000
Update seismic hazard maps for ground shaking	1-20	192,000	9,600	9,600	9,600
Develop predictive models for ground deformation	1-3	1,500	300	0	0
Develop seismic hazard maps for liquefaction	1-20	16,000	800	800	800
Develop seismic hazard maps for surface fault rupture	1-20	4,000	200	200	200
Develop seismic hazard maps for landslide potential	1-20	10,000	500	500	500
Total Cost		946,000	42,340	43,230	37,442

TABLE E.2 Cost Breakdown (in $million) and Scheduling for Task 9: Post-Earthquake Information Management

Task Breakdown	Task Total	Years 1-2 (Annualized)	Years 3-4 (Annualized)	Years 5-6 (Annualized)	Years 7-10 (Annualized)	Years 11-20 (Annualized)
PHASE 1						
Program management, travel and support[a]	1.27	0.635	—	—	—	—
Equipment & commercial software licensing	0.26	0.13	—	—	—	—
PHASE II						
Pilot projects (7 to 9 total)[b]	3.7	0	0.37	0.74	0.37	—
Operation costs[c]	9.4	0	0.6	0.6	0.6	0.46
Total Cost	**14.63**					

[a] Project management, PIMS system programmers, and testing/documentation staff would be budgeted at the GS13 salary range, about $125,000/year including benefits. The initial GS13 staff would be 4.5 in number (1 Project Lead at 0.5 FTE, 3 FTE system/software programmers, and 1 FTE testing/documentation staff). Administrative assistance and systems management would be budgeted at about $70,000/year including benefits, and would require 1 FTE (4.5 * $125,000/year + 1 * $70,000/year = 562,500 + 70,000 = 632,500).

[b] Each pilot project is estimated to have a development phase with 1 FTE at $90,000/year and an implementation phase with 1 to 2 FTEs at a cost of $140,000 to $420,000 per year, for a total cost of $230,000 to $510,000 per pilot project. For budgeting, an average of $370,000 per pilot project is assumed.

[c] Operation costs include project management, systems management, administrative assistance, user support, and maintenance programmers. It also includes data expansion and travel. All total, operation costs are estimated at $600,000 per year during Phase II, and $460,000 per year after Phase II.

TABLE E.3 Task Breakdown and Scheduling for Task 11: Observatory Network on Community Resilience and Vulnerability

Year	Annual Cost	Unit Cost	Cost Explanation
Year 1	$1.8 million	(a) new location nodes: $400k/yr (b) existing centers: $200k/yr	Annual cost assumes 6 nodes established in Phase I (3 at new locations, 3 at existing research centers). New nodes: personnel and infrastructure cost to establish node functions; initiate data collection. Existing centers: Less costly, as would build on existing personnel/infrastructure.
Year 2	$2.2 million	(a) and (b): Same as Year 1 (c) network coordinating grant: $400k/yr	6 nodes continue. Network coordinating funds (NCF) to facilitate and institutionalize coordination functions (e.g., measurement protocols, data archiving, network website, workshops, etc.).
Year 3	$3.8 million	(a), (b), (c): Same as above	6 nodes and NCF continue. Add 4 new nodes, potentially including "living laboratory" nodes.
Year 4	$3.8 million	(a), (b), (c): Same as above	10 nodes and NCF continue.
Year 5	$2.85 million	(a): $300k/yr (b): $150k/yr (c): $300k/yr	10 nodes and NCF continue. Cost reductions as infrastructure now set up and most urgent network actions completed (e.g., common measurement protocols, data archiving, etc.).
5-year TOTAL	$14.45 million		Sum of Years 1–5 costs.
Years 6–20	$2.85 million/yr		Maintain Year 5 funding levels; some nodes may end and be replaced by others through competition.

TABLE E.4 Summary Cost Breakdown (in $000) for Task 13: Techniques for Evaluation and Retrofit of Existing Buildings

Research and Development Task	Task Total ($)	Years 1-5 (Annualized) ($)	Years 6-10 (Annualized) ($)	Years 11-20 (Annualized) ($)
Program coordination and management	90,595	4,530	4,530	4,530
Establish a coordinated research program on existing buildings	1,200	60	60	60
Develop fragility and consequence functions for archaic components	5,875	858	218	50
Develop reliable tools for collapse computations	37,250	1,050	4,075	1,163
Large-scale laboratory testing of existing building systems, incl. improved component models	42,300	2,115	2,115	2,115
In-situ testing of existing buildings and components	109,000	50	9,750	6,000
Soil-structure interaction studies	21,500	325	1,045	1,465
Develop and deploy efficient retrofit methods/techniques	15,750	825	775	775
Develop and deploy techniques for NDE of existing construction and conditions	9,750	525	475	475
Develop and deploy a building rating system	4,000	700	0	50
Evaluate reliability of and update ASCE 41 procedures for PBD of existing buildings	18,650	2,065	1,665	0
Collect, curate, and archive building inventory data across the nation	135,650	6,880	6,750	6,750
Performance-based retrofit of nonstructural components and systems	875	175	0	0
Carbon footprint of retrofit building construction	775	155	0	0
Implementation: updating of standards and guidelines; risk reduction programs	50,400	2,580	2,500	2,500
Total Cost	**543,570**	**22,892**	**33,957**	**25,932**

TABLE E.5 Detailed Task Breakdown (in $000) and Scheduling for Task 13: Techniques for Evaluation and Retrofit of Existing Buildings[a]

Research and Development Task	Year	Task Breakdown ($)	Years 1-5 (Annualized) ($)	Years 6-10 (Annualized) ($)	Years 11-20 (Annualized) ($)
Program coordination and management	1-20	90,595	4,530	4,530	4,530
Establish a coordinated research program on existing buildings		1,200	60	60	60
Scoping studies and workshops	1,6,11,15	800	40	40	40
Development/update of work-plans	1,6,11,15	400	20	20	20
Develop fragility and consequence functions for archaic components (critical missing pieces)		5,875	858	218	50
Scoping studies and workshop	1	100	20	0	0
Development of a work-plan (not covered elsewhere, mining of existing data)	1	100	20	0	0
Experimentation using NEES facilities (Use experimental data generated elsewhere)	2-4	3,000	600	0	0
Numerical studies using improved hysteretic models developed elsewhere	4-6	1,800	180	180	0
Develop and document fragility and consequence functions	2-8	125	13	13	0
Update functions in Years 11-20	1-5	250	0	0	25
Synthesis of results and preparation of technical briefs	5 yearly	500	25	25	25
Develop reliable tools for collapse computations		37,250	1,050	4,075	1,163
Scoping studies and workshop	3	150	30	0	0
Development of work-plan (using also work on improved hysteretic models)	3	100	20	0	0

Experimentation using NEES facilities and E-Defense on multiple framing systems to collapse	6-10	12,000	0	2,400	0
Experimentation using NEES facilities on critical components of framing systems	4-7	7,500	750	750	0
Improved hysteretic models of structural components through failure	4-20	4,500	225	225	225
Understanding the triggers for collapse of framing systems	6-10	2,250	0	450	0
Improved system-level collapse computations and FE codes	6-15	2,250	0	225	113
Validation of improved computational procedures using NEES facilities and E-Defense	11-20	8,000	0	0	800
Synthesis of results and preparation of technical briefs	5 yearly	500	25	25	25
Large-scale laboratory testing of existing building systems, including improved component models		42,300	2,115	2,115	2,115
Scoping studies and workshop	1,6,11,15	600	30	30	30
Development of work-plan	1,6,11,15	600	30	30	30
Component testing program (NEES facilities): archaic and retrofitted	1-20	15,000	750	750	750
Systems testing program (NEES/E-Defense facilities): archaic and retrofitted	1-20	8,000	400	400	400
Develop nonlinear hysteretic models	1-20	9,000	450	450	450
Validate nonlinear hysteretic models	1-20	8,000	400	400	400
Develop guidelines and tools for FE analysis	5 yearly	600	30	30	30
Synthesis of data and preparation of a technical brief	5 yearly	500	25	25	25
In-situ testing of existing buildings and components		109,000	50	9,750	6,000
Scoping studies and workshop	3	150	30	0	0

continued

TABLE E.5 Continued

Research and Development Task	Year	Task Breakdown ($)	Years 1-5 (Annualized) ($)	Years 6-10 (Annualized) ($)	Years 11-20 (Annualized) ($)
Development of work-plan	3	100	20	0	0
Systems-level dynamic testing using NEES equipment: archaic and retrofitted	6-15	15,000	0	1,500	750
Numerical studies using systems-level dynamic test data	6-18	9,000	0	600	600
Systems-level testing to collapse: archaic and retrofitted	6-15	20,000	0	2,000	1,000
Numerical studies using collapse test data (supplement to above)	6-18	6,000	0	400	400
Component-level testing to failure: archaic and retrofitted	6-15	40,000	0	4,000	2,000
Numerical studies using component test data	6-18	18,000	0	1,200	1,200
Develop and validate nonlinear hysteretic models (included elsewhere)		0	0	0	0
Synthesis of data and preparation of technical briefs	10, 15, 20	750	0	50	50
Soil-structure interaction studies		21,500	325	1,045	1,465
Scoping studies and workshop	5	200	40	0	0
Development of work-plan	5	100	20	0	0
Centrifuge testing programs (alternate soils, layers, ground water table)	6-10	3,750	0	750	0
Develop simplified guidelines and tools for isolated structures	11-15	1,350	0	0	135
Develop simplified guidelines and tools for clusters of structures	11-15	1,350	0	0	135
Develop procedures for time-domain FE analysis	6-10, 15-20	2,700	0	270	135

Develop procedures for probabilistic SSI analysis	1-5	1,200	240	0	0
Implementation of time and frequency domain algorithms in FE codes	11-15	1,350	0	0	135
Validation of numerical tools by experimentation using NEES facilities and E-Defense	11-15	6,000	0	0	600
Update of tools and procedures in Years 16-20		3,000	0	0	300
Synthesis of data and preparation of technical briefs	5, 10, 15, 20	500	25	25	25
Develop and deploy efficient retrofit methods/techniques		15,750	825	775	775
Scoping studies and workshop	1	150	30	0	0
Development of work-plan	2	100	20	0	0
Develop alternate retrofit strategies	2-18	15,000	750	750	750
Deploy and test retrofit strategies (Included elsewhere)		0	0	0	0
Develop and validate nonlinear hysteretic models of retrofitted components (Included elsewhere)		0	0	0	0
Synthesis of data and preparation of technical briefs	5, 10, 15, 20	500	25	25	25
Develop and deploy techniques for NDE of existing construction and conditions		9,750	525	475	475
Scoping studies and workshop (utilize existing building construction; included elsewhere)		150	30	0	0
Development of work-plan	2	100	20	0	0
Develop alternate NDE strategies	3-20	9,000	450	450	450
Deploy and test alternate NDE strategies	3-20	0	0	0	0
Synthesis of data and preparation of technical briefs	5, 10, 15, 20	500	25	25	25

continued

TABLE E.5 Continued

Research and Development Task	Year	Task Breakdown ($)	Years 1-5 (Annualized) ($)	Years 6-10 (Annualized) ($)	Years 11-20 (Annualized) ($)
Develop and deploy a building rating system		4,000	700	0	50
Scoping studies and workshop	1	200	40	0	0
Development of work-plan	1	100	20	0	0
Numerical studies using data and PBD developed elsewhere	2-4	3,000	600	0	0
Update building rating system in Year 20	20	300	0	0	30
Synthesis of data and preparation of technical briefs	5 and 20	400	40	0	20
Evaluate reliability, and update, ASCE 41 procedures for performance-based design of existing buildings		18,650	2,065	1,665	0
Scoping studies and workshop	1	150	30	0	0
Development of work-plan	1	100	20	0	0
Develop a method to translate test data into acceptance criteria	2	200	40	0	0
Benchmark linear and nonlinear static procedures using nonlinear dynamic analysis	2-4	1,800	360	0	0
Benchmark all analysis procedures using earthquake data	2-4	1,800	360	0	0
Calibrate retrofit standards against performance expectations for new buildings	6-8	1,800	0	360	0
Clarify performance expectations in ASCE 31/41	9 and 10	250	0	50	0
Revise linear and nonlinear static procedures based on benchmarking	4-5	150	30	0	0
Evaluation of procedures/acceptance criteria using NEES facilities	2-5	6,000	1,200	0	0
Evaluation of system-level predictions using NEES facilities and E-Defense	6-8	6,000	0	1,200	0

Update of nonlinear dynamic analysis procedures	8-10	150	0	30	0
Synthesis of results and preparation of technical briefs	5 and 10	250	25	25	0
Collect, curate, and archive building inventory data across the nation		135,650	6,880	6,750	6,750
Scoping studies and workshop	1 task	200	40	0	0
Development of work-plan and standardized procedures	1 task	150	30	0	0
Develop procedures to track replacement of deficient buildings and update archive/loss estimates	2 and 3	300	60	0	0
50 cities	1-20	135,000	6,750	6,750	6,750
Performance-based retrofit of nonstructural components and systems		875	175	0	0
Scoping studies and workshop	1	150	30	0	0
Development of a work-plan	1	100	20	0	0
Develop procedures, tools, and recommendations for retrofit of architectural and M/E/P components and systems	2 and 3	500	100	0	0
Prepare technical brief	4	125	25	0	0
Carbon footprint of retrofit building construction		775	155	0	0
Scoping studies and workshop	1	150	30	0	0
Development of a work-plan	1	100	20	0	0
Carbon footprint calculation framework	2 and 3	100	20	0	0
Carbon footprint calculations for retrofit construction	3 and 4	100	20	0	0
Inclusion of carbon-based effects in loss computations	4 and 5	200	40	0	0
Prepare a technical brief	5	125	25	0	0

continued

TABLE E.5 Continued

Research and Development Task	Year	Task Breakdown ($)	Years 1-5 (Annualized) ($)	Years 6-10 (Annualized) ($)	Years 11-20 (Annualized) ($)
Implementation		50,400	2,580	2,500	2,500
Support updating of standards and guidelines		10,000	500	500	500
Develop methods to measure contributions of building stock to community resilience	1-3	400	80	0	0
Encourage risk reduction programs across the nation		40,000	2,000	2,000	2,000
Total Cost		543,570	22,892	33,957	25,932

[a] Bold headings within the Research and Development Task column represent the overarching title and cost summary of underlying non-bold components.

TABLE E.6 Summary Cost Breakdown (in $000) for Task 14: Performance-based Earthquake Engineering for Buildings

Research and Development Task	Task Total ($)	Years 1-5 (Annualized) ($)	Years 6-10 (Annualized) ($)	Years 11-20 (Annualized) ($)
Program coordination and management	148,585	7,429	7,429	7,429
NEES maintenance and operation; new equipment	500,000	25,000	25,000	25,000
Effect of ground deformation on buildings	8,975	250	895	325
Site response analysis	12,050	1,615	385	205
Constitutive models for soils	17,150	1,490	1,440	250
Soil-foundation-structure interaction	21,500	325	1,045	1,465
Selection and scaling of earthquake ground motions	2,550	75	205	115
Improved hysteretic models of structural components	42,300	2,115	2,115	2,115
Evaluate reliability of ASCE 41 procedures for performance-based design	14,600	1,665	1,255	0
Develop reliable tools for collapse computations	37,250	1,050	4,075	1,163
Develop fragility and consequence functions for modern and archaic components	5,875	858	218	50
Loss estimation tools for PBEE	925	0	0	93
Expected performance of code-conforming structures	3,450	690	0	0
Expand ATC-58 performance-based design methodology	2,800	280	150	65
Performance-based design of nonstructural components and systems	1,325	265	0	0
Smart/innovative/adaptive/sustainable components and framing systems	51,500	2,500	2,500	2,650
Carbon footprint of new and retrofit building construction	675	135	0	0
Implementation: Updating of Standards and Guidelines	20,000	1,000	1,000	1,000
Total Cost	**891,510**	**46,742**	**47,712**	**41,924**

TABLE E.7 Detailed Task Breakdown (in $000) and Scheduling for Task 14: Performance-based Earthquake Engineering for Buildings[a]

Research and Development Task	Year	Task Breakdown ($)	Years 1-5 (Annualized) ($)	Years 6-10 (Annualized) ($)	Years 11-20 (Annualized) ($)
Program coordination and management	1-20	148,585	7,429	7,429	7,429
NEES maintenance and operation; new equipment (contributes to multiple tasks)	1-20	500,000	25,000	25,000	25,000
Effect of ground deformation on buildings		8,975	250	895	325
Scoping studies and workshop	5	125	25	0	0
Development of work-plan	5	100	20	0	0
Experimental studies using NEES facilities	6-9	3,000		600	0
Effect of ground deformation on buildings	7-10	900		180	0
Validation of numerical tools by experimentation using NEES facilities and E-Defense	11-15	3,000		0	300
Techniques to mitigate the effects of liquefaction using NEES facilities	2-10	1,350	180	90	0
Synthesis of data and preparation of technical briefs	5 (ground deformation predictions), 10, 15, 20	500	25	25	25
Site response analysis		12,050	1,615	385	205
Scoping studies and workshop	1	150	30	0	0
Development of workplan	1	100	20	0	0
Field testing using NEES facilities (3 yrs of testing across the nation; incl. borelogs)	2-4	5,000	1,000	0	0

WUS site class coefficients	3-5	1,350	270	0	0
PNW site class coefficients	4-6	900	120	60	0
CEUS site class coefficients	5-7	2,250	150	300	0
Updates to work in years 11-20	11-20	1,800	0	0	180
Synthesis of data and preparation of a technical brief	5, 10, 15, 20	500	25	25	25
Constitutive models for soils		17,150	1,490	1,440	250
Scoping studies and workshop	1	150	30	0	0
Development of work-plan	1	100	20	0	0
Component testing program (small scale)	2-10	5,000	500	500	0
System testing program (larger-scale, laminar boxes, field testing) using NEES facilities	2-10	3,750	375	375	0
Develop equivalent linear models	2-10	2,250	225	225	0
Develop nonlinear hysteretic models	2-10	2,250	225	225	0
Implementation in FE codes	2-10	900	90	90	0
Updates to work in Years 11-20	11-20	2,250	0	0	225
Synthesis of data and preparation of a technical brief	5, 10, 15, 20	500	25	25	25
Soil-foundation-structure interaction		21,500	325	1,045	1,465
Scoping studies and workshop	5	200	40	0	0
Development of work-plan	5	100	20	0	0
Centrifuge testing programs (alternate soils, layers, ground water table)	6-10	3,750	0	750	0
Develop simplified guidelines and tools for isolated structures	11-15	1,350	0	0	135

continued

TABLE E.7 Continued

Research and Development Task	Year	Task Breakdown ($)	Years 1-5 (Annualized) ($)	Years 6-10 (Annualized) ($)	Years 11-20 (Annualized) ($)
Develop simplified guidelines and tools for clusters of structures	11-15	1,350	0	0	135
Develop procedures for time-domain FE analysis	6-10, 15-20	2,700	0	270	135
Develop procedures for probabilistic SSI analysis	1-5	1,200	240	0	0
Implementation of time and frequency domain algorithms in FE codes	11-15	1,350	0	0	135
Validation of numerical tools by experimentation using NEES facilities and E-Defense	11-15	6,000	0	0	600
Update of tools and procedures in Years 16-20	5, 10, 15, 20	3,000	0	0	300
Synthesis of data and preparation of technical briefs	5, 10, 15, 20	500	25	25	25
Selection and scaling of earthquake ground motions		2,550	75	205	115
Scoping studies and workshop (assumes that Phase 1 procedures for buildings developed under existing contracts)	2	150	30	0	0
Development of work-plan	2	100	20	0	0
Update of procedures for buildings, accounting for SSI effects	8-10	900	0	180	0
Update of procedures in Years 11-20		900	0	0	90
Synthesis of data and preparation of technical briefs	5, 10, 15, 20	500	25	25	25

Task	Year				
Improved hysteretic models of structural components		42,300	2,115	2,115	2,115
Scoping studies and workshop	1,6,11,15	600	30	30	30
Development of work-plan	1,6,11,15	600	30	30	30
Component testing program (NEES facilities)	1-20	15,000	750	750	750
Systems testing program (NEES/E-Defense facilities)	1-20	8,000	400	400	400
Develop nonlinear hysteretic models	1-20	9,000	450	450	450
Validate nonlinear hysteretic models	1-20	8,000	400	400	400
Develop guidelines and tools for FE analysis	Every 5 years	600	30	30	30
Synthesis of data and preparation of a technical brief	Every 5 years	500	25	25	25
Evaluate reliability of ASCE 41 procedures for PBD		14,600	1,665	1,255	0
Scoping studies and workshop	1	150	30	0	0
Development of work-plan	1	100	20	0	0
Benchmark linear and nonlinear static procedures using nonlinear dynamic analysis	2-4	1,800	360	0	0
Revise linear and nonlinear static procedures based on benchmarking	4-5	150	30	0	0
Evaluation of procedures/acceptance criteria using NEES facilities	2-5	6,000	1,200	0	0
Evaluation of system-level predictions using NEES facilities and E-Defense	6-8	6,000	0	1,200	0
Update of nonlinear dynamic analysis procedures	8-10	150	0	30	0
Synthesis of results and preparation of technical briefs	5 and 10	250	25	25	0
Develop reliable tools for collapse computations		37,250	1,050	4,075	1,163
Scoping studies and workshop	3	150	30	0	0

continued

TABLE E.7 Continued

Research and Development Task	Year	Task Breakdown ($)	Years 1-5 (Annualized) ($)	Years 6-10 (Annualized) ($)	Years 11-20 (Annualized) ($)
Development of work-plan	3	100	20	0	0
Experimentation using NEES facilities and E-Defense on multiple framing systems to collapse	6-10	12,000	0	2,400	0
Experimentation using NEES facilities on critical components of framing systems	4-7	7,500	750	750	0
Improved hysteretic models of structural components through failure (using also work on improved hysteretic models)	4-20	4,500	225	225	225
Understanding the triggers for collapse of multiple framing systems	6-10	2,250	0	450	0
Improved system-level collapse computations and FE codes	6-15	2,250	0	225	113
Validation of improved computational procedures using NEES facilities and E-Defense	11-20	8,000	0	0	800
Synthesis of results and preparation of technical briefs	Every 5 years	500	25	25	25
Develop fragility and consequence functions for modern and archaic components		5,875	858	218	50
Scoping studies and workshop	1	100	20	0	0
Development of a workplan	1	100	20	0	0

Task	Year				
Experimentation using NEES facilities	2-4 (critical missing pieces)	3,000	600	0	0
Numerical studies using improved hysteretic models developed elsewhere	4-6	1,800	180	180	0
Develop and document fragility and consequence functions (not covered elsewhere, mining of existing data)	2-8	125	13	13	0
Update functions in Years 11-20 (Use experimental data generated elsewhere)		250	0	0	25
Synthesis of results and preparation of technical briefs	5-yearly	500	25	25	25
Loss estimation tools		925	0	0	93
Scoping studies and workshop	11	100	0	0	10
Development of a work-plan	11	100	0	0	10
Develop and implement tools for ground deformation	12-13	200	0	0	20
Develop and implement tools for fire-following-earthquake	13-14	200	0	0	20
Develop and implement tools related to carbon emissions	12-13	200	0	0	20
Synthesis of results and preparation of technical briefs	15	125	0	0	13
Expected performance of code-conforming structures		3,450	690	0	0
Scoping studies and workshop	1	150	30	0	0
Development of a work-plan	1	100	20	0	0
Assess performance of code-conforming structures—WUS	2-3	500	100	0	0
Assess performance of code-conforming structures—PNW	2-3	500	100	0	0
Assess performance of code-conforming structures—CEUS	2-3	1,000	200	0	0
Revise ASCE-7 provisions, values of R, etc	4 and 5	1,000	200	0	0
Synthesis of results and preparation of technical briefs	5	200	40	0	0

continued

TABLE E.7 Continued

Research and Development Task	Year	Task Breakdown ($)	Years 1-5 (Annualized) ($)	Years 6-10 (Annualized) ($)	Years 11-20 (Annualized) ($)
Expand ATC-58 PBD methodology		2,800	280	150	65
Scoping studies and workshop	1	150	30	0	0
Development of a work-plan	1	100	20	0	0
Expand methodology to include ground deformation	11-13	200	0	0	20
Expand methodology to include post-earthquake flooding	11-13	200	0	0	20
Extend methodology to lifelines	2-4	500	100	0	0
Extend methodology to earthen structures	2-4	200	40	0	0
Extend methodology to selected infrastructure	6-10	500	0	100	0
Extend methodology to flood protection structures	2-4	200	40	0	0
Prepare technical briefs	5, 10, 15	750	50	50	25
PB design of nonstructural components and systems		1,325	265	0	0
Scoping studies and workshop	1	100	20	0	0
Development of a work-plan	1	100	20	0	0
Develop procedures and tools for architectural and M/E/P components systems	2-5	1,000	200	0	0
Prepare technical brief	5	125	25	0	0
Smart/innovative/adaptive/sustainable components and framing systems		51,500	2,500	2,500	2,650
Development and deployment of smart framing systems, incl. hysteretic models	1-20	20,000	1,000	1,000	1,000

Development and deployment of adaptive components, incl. hysteretic models	1-20	20,000	1,000	1,000	1,000
Development and deployment of sustainable components (systems), incl. hysteretic models	1-20	10,000	500	500	500
Preparation of standards and guidelines for smart framing systems	11-20	500	0	0	50
Preparation of standards and guidelines for adaptive components	11-20	500	0	0	50
Preparation of standards and guidelines for sustainable components	11-20	500	0	0	50
Carbon footprint of new and retrofit building construction		675	135	0	0
Scoping studies and workshop	1	150	30	0	0
Development of a work-plan	1	100	20	0	0
Carbon footprint calculation framework for new and archaic framing systems	2 and 3	100	20	0	0
Carbon footprint calculations for retrofit construction	3 and 4	100	20	0	0
Inclusion of carbon-based effects in loss computations	4 and 5	100	20	0	0
Prepare a technical brief	5	125	25	0	0
Implementation: updating of standards and guidelines		20,000	1,000	1,000	1,000
Total Cost		891,510	46,742	47,712	41,924

[a] Bold headings within the Research and Development Task column represent the overarching title and cost summary of underlying non-bold components.

TABLE E.8 Summary Cost Breakdown (in $000) for Task 16: Next Generation Sustainable Materials, Components, and Systems

Research and Development Task	Task Total ($)	Years 1-5 (Annualized) ($)	Years 6-10 (Annualized) ($)	Years 11-20 (Annualized) ($)
Engineering research center management	55,735	2,787	2,787	2,787
Investigate and characterize new materials	73,300	4,685	6,575	1,700
Devise new modular precast components and framing systems	8,175	0	835	400
Develop tools, technologies, and details to join new materials	16,000	0	1,100	1,050
Prototype components, connections, and framing systems	8,200	0	469	586
Moderate and full-scale testing of components with new materials using NEES infrastructure	50,700	0	0	5,070
Full-scale tests of 3D framing systems	15,550	0	0	1,555
Develop design tools and equations for new materials	8,000	0	0	800
Develop and characterize a new family of adaptive materials	15,650	0	1,600	765
Develop robust algorithms for controlling the response of adaptive materials	5,300	0	0	530
Develop a family of low-cost, low-power, zero maintenance wireless sensors	12,800	685	625	625
Prototype adaptive materials and components at the macro scale	8,100	0	0	810
Develop algorithms to control response of framing systems with adaptive components	8,000	0	0	800
Moderate and full-scale testing of adaptive components using NEES infrastructure	25,550	0	0	2,555
Full-scale tests of adaptive 3D framing systems	15,350	0	0	1,535
Develop design tools and equations for adaptive components and systems	8,000	0	0	800
Total Cost	334,410	8,157	13,990	22,368

TABLE E.9 Summary Cost Breakdown (in $000) for Task 16: Next Generation Sustainable Materials, Components, and Systems[a]

Research and Development Task	Year	Task Breakdown ($)	Years 1-5 (Annualized) ($)	Years 6-10 (Annualized) ($)	Years 11-20 (Annualized) ($)
Engineering research center management	1-20	55,735	2,787	2,787	2,787
Investigate and characterize new materials		73,300	4,685	6,575	1,700
Scoping studies and workshop	1	200	40	0	0
Development of work-plan	1	100	20	0	0
Small-scale characterization studies—concrete (low-cement, very high strength, fiber-reinforced)	2-10	15,000	1,500	1,500	0
Small-scale characterization studies—metals	2-10	15,000	1,500	1,500	0
Small-scale characterization—polymers	2-10	15,000	1,500	1,500	0
Small-scale characterization—others	6-15	7,500	0	750	375
Develop micro-mechanical models for new materials	6-20	18,000	0	1,200	1,200
Synthesis of data and preparation of technical briefs	5, 10, 15, 20	2,500	125	125	125
Devise new modular precast components and framing systems		8,175	0	835	400
Scoping studies and workshop	6	200	0	40	0
Development of work-plan	6	100	0	20	0
Develop new components and systems	7-15	7,500	0	750	375
Synthesis of data and preparation of a technical brief	10, 15, 20	375	0	25	25
Develop tools, technologies, and details to join new materials		16,000	0	1,100	1,050
Scoping studies and workshop	6	150	0	30	0
Development of work-plan	6	100	0	20	0

continued

TABLE E.9 Continued

Research and Development Task	Year	Task Breakdown ($)	Years 1-5 (Annualized) ($)	Years 6-10 (Annualized) ($)	Years 11-20 (Annualized) ($)
Component testing program (small scale)	7-20	15,000	0	1,000	1,000
Synthesis of data and preparation of a technical brief	10, 15, 20	750	0	50	50
Prototype components, connections, and framing systems		8,200	0	469	586
Scoping studies and workshop	8	100	0	20	0
Development of work-plan	8	100	0	20	0
Prototype components, connections, and systems	9-15	7,500	0	429	536
Synthesis of data and preparation of a technical brief	15, 20	500	0	0	50
Moderate and full-scale testing of components with new materials using NEES infrastructure		50,700	0	0	5,070
Scoping studies and workshop	11	100	0	0	10
Development of work-plan	11	100	0	0	10
Component testing program (reaction walls/floors, laminar boxes) using NEES facilities	12-20	30,000	0	0	3,000
Develop nonlinear hysteretic models and design equations for materials standards	12-20	9,000	0	0	900
Implementation of models in FE codes	12-20	9,000	0	0	900
Develop fragility and consequence functions for PBD	12-20	2,000	0	0	200
Synthesis of data and preparation of a technical brief	15, 20	500	0	0	50
Full-scale tests of 3D framing systems		15,550	0	0	1,555
Scoping studies and workshop	16	200	0	0	20
Development of work-plan	16	100	0	0	10

Full-scale testing using NEES facilities and E-Defense	17-20	12,500	0	0	1,250
Validation of numerical tools and models	18-20	2,250	0	0	225
Synthesis of data and preparation of technical briefs	15 and 20	500	0	0	50
Develop design tools and equations for new materials		8,000	0	0	800
Scoping studies and workshop	16	200	0	0	20
Development of work-plan	16	100	0	0	10
Develop design tools and equations	17-20	4,500	0	0	450
Prepare materials standards (e.g., ACI 318)	17-20	2,700	0	0	270
Synthesis of data and preparation of technical briefs	20	500	0	0	50
Develop and characterize a new family of adaptive materials		15,650	0	1,600	765
Scoping studies and workshop	6	200	0	40	0
Development of work-plan	6	150	0	30	0
Develop and characterize new materials and fluids	7-15	15,000	0	1,500	750
Synthesis of data and preparation of technical briefs	10 and 15	300	0	30	15
Develop robust algorithms for controlling the response of adaptive materials		5,300	0	0	530
Scoping studies and workshop	11	150	0	0	15
Development of work-plan	11	150	0	0	15
Algorithm development and validation	12-18	4,500	0	0	450
Synthesis of data and preparation of a technical brief	15 and 20	500	0	0	50
Develop a family of low-cost, low-power, zero maintenance wireless sensors		12,800	685	625	625
Scoping studies and workshop	1	200	40	0	0
Development of work-plan	1	100	20	0	0
Develop sensors	2-20	12,000	600	600	600

continued

TABLE E.9 Continued

Research and Development Task	Year	Task Breakdown ($)	Years 1-5 (Annualized) ($)	Years 6-10 (Annualized) ($)	Years 11-20 (Annualized) ($)
Synthesis of results and preparation of technical briefs	5, 10, 15, 20	500	25	25	25
Prototype adaptive materials and components at the macro scale		8,100	0	0	810
Scoping studies and workshop	11	200	0	0	20
Development of work-plan	11	150	0	0	15
Prototype components, connections, and systems	12-15	7,500	0	0	750
Synthesis of results and preparation of technical briefs	15 and 20	250	0	0	25
Develop algorithms to control response of framing systems with adaptive components		8,000	0	0	800
Scoping studies and workshop	11	150	0	0	15
Development of a work-plan	11	100	0	0	10
Algorithm development and validation by testing	12-20	7,500	0	0	750
Synthesis of results and preparation of technical briefs	15 and 20	250	0	0	25
Moderate and full-scale testing of adaptive components using NEES infrastructure		25,550	0	0	2,555
Scoping studies and workshop	15	200	0	0	20
Development of work-plan	15	100	0	0	10
Component testing program (reaction walls/floors, laminar boxes) using NEES facilities	16-20	15,000	0	0	1,500
Develop nonlinear hysteretic models and design equations for materials standards	16-20	4,500	0	0	450
Implementation of models in FE codes	16-20	4,500	0	0	450

Develop fragility and consequence functions for PBD	16-20	1,000	0	0	100
Synthesis of data and preparation of a technical brief	20	250	0	0	25
Full-scale tests of adaptive 3D framing systems		15,350	0	0	1,535
Scoping studies and workshop	16	200	0	0	20
Development of work-plan	16	150	0	0	15
Full-scale testing using NEES facilities and E-Defense	17-20	12,500	0	0	1,250
Validation of numerical tools and models	18-20	2,250	0	0	225
Synthesis of data and preparation of technical briefs	20	250	0	0	25
Develop design tools and equations for adaptive components and systems		8,000	0	0	800
Scoping studies and workshop	16	200	0	0	20
Development of work-plan	16	100	0	0	10
Develop design tools and equations	17-20	4,500	0	0	450
Prepare materials standards (e.g., ACI 318)	17-20	2,700	0	0	270
Synthesis of data and preparation of technical briefs	20	500	0	0	50
Total Cost		334,410	8,157	13,990	22,367

[a] Bold headings within the Research and Development Task column represent the overarching title and cost summary of underlying non-bold components.

TABLE E.10 Cost Breakdown (in $million) for Task 18: Earthquake-Resilient Communities and Regional Demonstration Projects[a]

Budget Component	Task Total Cost	Expenditure by Year													
		1-2	3	4	5	6	7	8-9	10	11-13	14	15-17	18	19	20
Program management and support	22	0.5	0.5	0.88	0.88	0.88	0.88	0.88	1.38	1.38	1.38	1.38	1.38	1.38	1.375
National outreach and information	20	1.0	1.0	1.0	1.0	1.0	1.0	1.0	1.0	1.0	1.0	1.0	1.0	1.0	1.0
Strategy preparation	1	0.5	0	0	0	0	0	0	0	0	0	0	0	0	0
Data collection and profile	5	0.25	0.25	0.25	0.25	0.25	0.25	0.25	0.25	0.25	0.25	0.25	0.25	0.25	0.25
Independent research	20	1.0	1.0	1.0	1.0	1.0	1.0	1.0	1.0	1.0	1.0	1.0	1.0	1.0	1.0
Community outreach	840	0	7.5	15	30	45	52.5	60	60	60	60	60	45	30	15
Gap-filling (tools, guidance, etc.)	5	0.25	0.25	0.25	0.25	0.25	0.25	0.25	0.25	0.25	0.25	0.25	0.25	0.25	0.25
Monitor, evaluate, analysis, feedback, revision	0	0.25	0.5	0.5	0.5	0.5	0.5	0.5	0.5	0.5	0.5	0.5	0.5	0.5	0.5
Revise strategy	2	0	0	0	0	0.5	0	0	0.5	0	0.5	0	0.5	0	0
Annual workshop	3	0.15	0.15	0.15	0.15	0.15	0.15	0.15	0.15	0.15	0.15	0.15	0.15	0.15	0.15
State participation (estimate 30 states)	73.5	0	0.75	2.25	3.0	4.5	4.5	4.5	4.5	4.5	4.5	4.5	4.5	4.5	4.5
Total Cost	1,001	3.9	12	21.3	37	54	61	68.5	69.5	69	69.5	69	54.5	39	24

^{a} Task 18 Assumptions:

• Community component (ramping up and down)—The proposal is to identify early adoption communities to conduct pilot and demonstration programs in about 280 communities. The project would begin by selecting 10 pilot communities in years 3 and 4 to test and develop the program and materials. In years 5-16 the project would add 20 demonstration communities. Pilot and demonstration communities would receive an average of \$750,000 per year for 4 years after which they would be expected to have the information, understanding, tools, organizations, motivation, and support needed to sustain resilience-improving efforts into the future.

• Data collection (data profiling) would occur during the first 2 years in coordination with development of the strategy, and then each year as new communities require individual hazard assessments, etc.

• Program management and support staff costs would be budgeted at the GS13 salary range, about \$125,000/year including benefits. The initial staff would be 4 in number, 3 additional staff persons would be added in year 4, and 3 additional staff persons in year 10 for a total of 11. (4×125 = 500,000; $7 \times .125 = .875$; $11 \times .125 = 1.375$)

• Existing centers at Delaware (DRC), University of South Carolina, Texas A&M, UCLA, USC CREATES Center have the capability to do community assessments and evaluation of the program and individual communities.

• National outreach provides funds for providing information, engaging specialists, evaluating events that occur, etc.

• Consultation at the state level will be important in the selection and support of the communities.

• Leverage with the other components—National Seismic Hazard Maps, HAZUS, community-specific hazard maps, earthquake provisions in codes, training materials for professionals, learning from earthquakes presentations at annual workshops, NSF could sponsor research on the process.

• Must describe outcomes of more resilient communities in more meaningful terms such as codes adopted and enforced, community understanding, land-use plans that consider earthquake hazard, that measures taken before earthquakes will have a multihazard application

• Community strategy could "require" a comprehensive school earthquake safety element (awareness, curriculum, preparedness, building analysis and mitigation).

Appendix F

Acronyms and Abbreviations

AEL	Annualized Earthquake Loss
AELR	Annualized Earthquake Loss Ratios
ALA	American Lifelines Alliance
ANSS	Advanced National Seismic System
ARRA	American Recovery and Reinvestment Act
ASCE	American Society of Civil Engineers
ATC	Applied Technology Council
BSSC	Building Seismic Safety Council
CARRI	Community and Regional Resilience Institute
CDMS	Comprehensive Data Management System
CGE	Computable General Equilibrium
CISN	California Integrated Seismic Network
CLEANER	Collaborative Large-Scale Engineering Analysis Network for Environmental Research
CREW	Cascadia Region Earthquake Workgroup
CUREE	Consortium of Universities for Research in Earthquake Engineering
CUSEC	Central U.S. Earthquake Consortium
DELM	Direct Economic Loss Modules
DHS	Department of Homeland Security
DOGAMI	Oregon Department of Geology and Mineral Industries

DRC	Southwest Indiana Disaster Resistant Community Corporation
DSER	direct static economic resilience
EERI	Earthquake Engineering Research Institute
EEW	earthquake early warning
FEMA	Federal Emergency Management Agency
GEER	Geotechnical Extreme Event Reconnaissance Association
GIS	Geographic Information System
GPS	Global Positioning System
HAZUS	FEMA's Hazards U.S.
HAZUS-MH	Hazards U.S.-Multi-Hazard
HUGs	HAZUS User Groups
I-O	Input-Output
IDFBS	Indiana Department of Fire and Building Services
IELM	Indirect Economic Loss Modules
IIPLR	Insurance Institute for Property Loss Reduction
InSAR	Interferometic Synthetic Aperture Radar
ISDR	The United Nations International Strategy for Disaster Reduction
LFE	Learning from Earthquakes program, managed by EERI
LiDAR	Light Detection and Ranging
LTER	Long-Term Ecological Research
MAE	Mid-America Earthquake Center
MCEER	Multidisciplinary Center for Earthquake Engineering Research
MP	Mathematical programming
NCSA	National Center for Supercomputer Applications
NEES	Network for Earthquake Engineering Simulation
NEHRP	National Earthquake Hazards Reduction Program
NEIC	The National Earthquake Information Center
NEON	National Ecological Observatory Network
NEPEC	National Earthquake Prediction Evaluation Council
NGA	Next Generation Attenuation
NHRAIC	Natural Hazards Research Applications and Information Center

NIBS	National Institute of Building Sciences
NIPP	Department of Homeland Security's Infrastructure Protection Program
NIST	National Institute of Standards and Technology
NRC	National Research Council
NSF	National Science Foundation
NSTC	National Science and Technology Council
OES	California Governor's Office of Emergency Services
OpenSees	Open System for Earthquake Engineering Simulation
PEER	Pacific Earthquake Engineering Research Center
PGA	Peak Ground Acceleration
PIMS	Post-earthquake Information Management System
R&D	research and development
RAVON	Resiliency and Vulnerability Observatory Network
ROVER	Rapid Observation of Vulnerability and Estimation of Risk
SEAW	Structural Engineers Association of Washington
SPUR	San Francisco Planning and Urban Research Association— Resilient City Initiative
STEP	Short-Term Earthquake Probability
TCLEE	Technical Council on Lifeline Earthquake Engineering
UCERF2	Uniform California Earthquake Rupture Forecast— Version 2
USGS	U.S. Geological Survey
URM	unreinforced masonry